## Scribe Publications
## CLIMATE CODE RED

David Spratt is a Melbourne businessman, climate-policy
analyst, and co-founder of Carbon Equity, which advocates
personal carbon allowances as the most fair and equitable
means of rapidly reducing carbon emissions. He has
extensive advocacy experience in the peace movement, and
in developing community-campaign communication and
marketing strategies.

Philip Sutton is the convener of the Greenleap Strategic
Institute, a non-profit environmental-strategy think tank
and advisory organisation promoting the very rapid
achievement of global and local ecological sustainability.
He is also the founder and director of strategy for Green
Innovations, and an occasional university lecturer on global
warming science and strategies for sustainability.

Philip has worked on a number of advisory and policy
committees for Australian state and federal governments,
was the architect of the Victorian Flora and Fauna Guarantee
Act, and is a former president of the Australian and New
Zealand Society for Ecological Economics (2001–2003).

*For Helen O'Shea*
– David

*To my children, Dan and Joey*
– Philip

# CLIMATE
# CODE RED

the case for emergency action

David Spratt
& Philip Sutton

**SCRIBE**
*Melbourne*

Scribe Publications Pty Ltd
PO Box 523
Carlton North, Victoria, Australia 3054
Email: info@scribepub.com.au

Published in Australia and New Zealand by Scribe 2008

Cover design by Tamsyn Hutchinson
Typeset in 11.5/15 pt Dante by the publishers
Printed and bound in Australia by Griffin Press
Only wood grown from sustainable regrowth forests is used in the
manufacture of paper found in this book.

National Library of Australia
Cataloguing-in-Publication data

Spratt, David.
Climate code red: the case for emergency action
Carlton North, Vic.: Scribe Publications, 2008.

9781921372209

Global warming; Global temperature changes.

363.73874

www.scribepublications.com.au

# Contents

WITHDRAWN

## Part Three: The Climate Emergency

*Note*

All temperatures cited in this book are in Celsius, and increases in temperature are from 1750 pre-industrial levels unless otherwise stated.

# Forewords

*Ian Dunlop*

As the world's population rises toward nine billion by mid-century, the inevitable logic of exponential growth in consumption is now hitting the real limits of global ecosystems and resource availability. The immediate pressure points are human-induced climate change, water availability, and peaking global oil supply, which are converging rapidly in a manner not previously experienced. But those pressure points constitute only the tip of the broader global-sustainability iceberg: further constraints and limits will become increasingly evident as the major developing countries move up the growth escalator.

This situation is not unexpected: it has been forecast for decades, going back before the 1972 publication of *The Limits to Growth*, a book that described how expanding human population and consumption patterns would run up against the limits of the natural world. In the meantime, we have created a political and capitalist system which has proved incapable of recognising that the most important factor for its own survival is the preservation of a global biosphere fit for human habitation. Our institutions are totally short-term focused: politically, due to the electoral cycle, and corporately, due to perverse incentives. Thus, we are uniquely ill-equipped to handle these major problems, which are all long-term.

Our ideological preoccupation with a market economy that is based on maximising short-run profit is rapidly leading us towards an uninhabitable planet. As inconvenient as it may be, politically and corporately, conventional economic growth and rampant consumerism cannot continue. Markets are important, but they operate within rules; henceforth, the rules must change to ensure long-run sustainability.

Nationalism and short-term vested interests have so far prevented the development of a global governance framework capable of handling this Tragedy of the Commons, and the issue of global sustainability is now much bigger than any nation state. Global warming, in particular, is moving far faster than scientists had predicted, to the point that we are already in the danger zone.

The stark fact is that we face a global sustainability emergency, but it is impossible to design realistic solutions unless we first understand and accept the size of the problem. We know those solutions; what is lacking is the political will, first, to honestly articulate the problem and, second, to implement those solutions.

Unadorned by political spin, *Climate Code Red* is a sober, balanced analysis of this challenge that proposes a realistic framework to tackle the emergency. It should be essential reading for all political and corporate leaders and, particularly, for the community. The extent of change that we require will only occur if the political and corporate world sees that the community is demanding it.

If we are to have a reasonable chance of maintaining a habitable planet, placing our efforts on an emergency footing is long overdue. We only play this game once; a trial run is not an option.

**Ian Dunlop** is a former international oil, gas, and coal industry executive. He chaired the Australian Coal Association from 1987–88 and the Australian Greenhouse Office Experts Group on Emissions Trading from 1998–2000, and he was CEO of the Australian Institute of Company Directors from 1997–2001. He is chairman of the Australian National Wildlife Collection Foundation (CSIRO) and deputy convenor of the Australian Association for the Study of Peak Oil.

## Ken Caldeira

For my PhD research, I studied what happened to ocean chemistry at the time the dinosaurs became extinct. The meteorite that destroyed the dinosaurs also acidified the oceans, leading to the disappearance of coral reefs and many other marine organisms. It is becoming clear that modern industrial civilisation is generating a new mass extinction (with its own ocean acidification) of a magnitude not seen since that destruction of the dinosaurs some 65 million years ago.

Over the past few centuries, vast natural ecosystems on land and in the water have been converted to human use and abuse. Our carbon dioxide emissions are heating the planet and acidifying the oceans. Our physical environment is changing at a rate that is faster than at any time in the past hundreds of millions of years, except for those rare cataclysmic events that have killed off most life on Earth. Spratt and Sutton point out that if 'business as usual' means losing Arctic ecosystems, losing coral reefs, altering the great weather patterns, and so on, then we simply cannot afford it—the cost is too great.

At the risk of oversimplification: we are forced to make a choice. Either we can decide to live in a 'wilderness world' in which we use our technology to minimise our environmental footprint, and we grow and develop in ways that are consistent with long-term flourishing of the rich diversity of life in this planet; or we can continue heading towards a nightmare vision of an Earth where climate is shifting and species are getting tossed overboard every day, every hour. (Invasive generalists, including weeds and the wealthy, may do fairly well, but specialist species and poor people have a threatened future ahead of them.) We will either learn to live with the world, or wreck it—and in wrecking the world, we will lose.

There is inertia both in the climate system and in our industrial infrastructure. Inertia in the climate system means we can pass thresholds now that set us on an irreversible trajectory to future tragedy. Inertia in our industrial infrastructure means that, under most accepted scenarios, without early retirement of major segments of our industrial capacity, it will take many decades to replace the coal-, oil-, and gas-burning devices that pervade our planet. 'Business as usual' is accelerating us into ever-greater environmental risk—and eventually, that risk will come home to roost. As Spratt and Sutton point out, 'business as usual' must end now, if we are to allow our children and ourselves a more natural world in which humans tread lightly and live well. Sensibly, *Climate Code Red* asks us to take stock of the climate and sustainability emergency that is unravelling around us and respond with a large-scale transition to a post-carbon economy. There is no time for slow transitions.

**Ken Caldeira** is director of the Caldeira Lab of the Department of Global Ecology at Stanford University's Carnegie Institution of Washington, whose research focuses on improving the science base needed to allow human civilisation to develop while protecting our environmental endowment. He also conducted research for the Energy and Environmental Sciences Directorate of the Lawrence Livermore National Laboratory from 1993 to 2005.

# A Lot More Trouble

'This is an emergency, and for emergency situations we
need emergency action ...'
— UN secretary-general Ban Ki-Moon, 10 November 2007

'I can't end this email without acknowledging that we are in
a lot more climate trouble than we thought.' This response
from a US-based polar researcher during a discussion we
were having about how quickly Greenland might melt is
not an orthodox scientific statement, but its disturbing tone
expresses a level of anxiety and honesty that we heard many
times while writing this book.

Perhaps the frankness of such responses reflected the
fact that we are not climate scientists, and that we were
asking questions not as peers but as policy researchers. In
conducting our enquiries, we conversed with and drew on the
work of many climate scientists who gave generously of their
time, patiently answering sometimes-wayward questions,
and welcoming our enquiries. Although our previous work

had included advocacy on environmental and community problems, we found the scope and depth of climate research, the nuances, and interpretative differences between scientists a challenge. Yet it is critical that non-scientists engage with the science if all of us are to plot a pathway to a safe climate.

Climate scientists generally work in a specialised field, and the release of their scientific results and projections incorporates assessements of risks, probabilities, and uncertainties that can lead them to feel reticent about commenting publicly on the broader aspects of global-warming impacts and policies. Those outside the research community, however, have a different vantage-point in viewing the disparate evidence, which may explain why some of the most compelling writing on global warming has come from writers such as George Monbiot, Fred Pearce, Mark Lynas, and Elizabeth Colbert.

*Climate Code Red* explores what 'a lot more climate trouble' means, why it differs from the public story, and how we should go about thinking of new solutions to this global emergency. It concludes that we must cast aside climate policies that are doomed to fail, and that we must act with foresight and courage, because our task is urgent.

The evidence we have gathered has convinced us that we have only one chance to solve the global warming problem. Just as in hospitals, where 'code red' denotes a patient who needs advanced life-support, the phrase signals an emergency: an alarm that rings now, for all life on this fragile planet.

Debate over climate change took a radical new turn in September 2007, when research data revealed that the floating sea-ice in the polar north was disintegrating at a frightening speed—in the words of Penn State University climatologist

Richard Alley, 'one hundred years ahead of schedule'. Eight million square kilometres of Arctic sea-ice is breaking up, and this demands that we look anew at the impact of global warming, and at what we must do to return to a safe-climate world.

Industrial activity is propelling the world's climate to a hot state not experienced for a million years, at a time long before modern humans evolved. We face a perilous journey across unfamiliar terrain, close to a precipice that, should we cross it, will see changes beyond recognition to life on Earth.

This is not an exaggerated claim; it is the sober view of many of the world's leading climate scientists, including NASA scientist Jay Zwally. When he was a young man, Zwally hauled coal for work. At the end of 2007, he told a gathering of fellow climate experts: 'The Arctic is often cited as the canary in the coalmine for climate warming ... and now as a sign of climate warming, the canary has died. It is time to start getting out of the coal mines.'

Robert Corell, chairman of the Arctic Climate Change Impact Assessment (ACCIA), is equally blunt:

> For the last 10,000 years we have been living in a remarkably stable climate that has allowed the whole of human development to take place ... Now we see the potential for sudden changes of between 2 and 6 degrees Celsius [by the end of this century].* We just don't know what the world is like at those temperatures. We are climbing rapidly out of mankind's safe zone into new territory, and we have no idea if we can live in it.

---

* The increase to date has been 0.8 degrees.

In the recent past, the story of climate change has been one of sudden and disruptive fluctuation as the Earth seesawed between ice ages and warm periods. This history warns that we must expect the unexpected, because dramatic changes that tip regional climates from one state to another can set off chains of events that echo around the globe.

Most of us think of climate change as a gradual, linear process that involves a smooth relationship between increasing levels of greenhouse gases and rising temperatures—that like the kitchen oven, if we are slowly turning up the control, we will produce a predictable warming. But climate doesn't work like that.

In fact, we live in a climate world of chaotic, non-linear transitions, where a small increase in the level of greenhouse gases, or in the energy imbalance of the climate system, can make a huge difference. An element of the climate system can flip from one state to another quickly and unpredictably. This is now occurring at the North Pole, where a tipping point, or critical threshold, has been passed, and an area of summer sea-ice once as large as Australia is disintegrating quickly.

Further south, if the changing climate were to produce four or five consecutive years of drought in the Amazon, it might become sufficiently dry for wildfires to destroy much of the rainforest and for burning carbon to pour into the skies. This change in the regional climate pattern would further reduce rainfall, and the drying and dead forest would release very large amounts of greenhouse gases. These impacts, like many others, would cause further threshold events.

If this kind of momentum builds sufficiently, and enough tipping points are crossed, we will pass a point of no return. We wish it were otherwise. Indeed, this is not a book we intended to write; but when our work led us to understand that we had already entered the era of dangerous climate change, it became a story we felt compelled to tell.

Originally, we wrote this book as a report to address three areas of climate policy that we wished to bring to the attention of the Garnaut Climate Change Review (the Garnaut Review): the implications of recent climate research, appropriate reduction targets, and the case for emergency action. The review was commissioned by the Australian federal Labor leader Kevin Rudd, and the state and territory Labor governments, six months prior to the November 2007 federal election that swept Labor to power.

Our first concern in presenting ideas to the review was to draw attention to the seriousness of recent climate data.

Our second concern was to show that a response to the climate crisis in 'politics as usual' mode would not be fast enough to solve the problem.

Our third concern was to encourage the review to choose targets that would achieve a safe result and not replicate mistakes made elsewhere in setting targets that would be ineffective. An example of this was the Stern Review, delivered to the UK government in late 2006 and received enthusiastically in most quarters. The economist Sir Nicholas Stern had graphically outlined the future impacts of rising global temperatures. An increase of 2 degrees above the pre-industrial level was not acceptable, he explained, because it would likely mean, amongst other things, the loss of 15–40 per cent of species, a loss of fresh water of 20–30 per cent in vulnerable regions, and the potential for the Greenland ice sheet to begin melting irreversibly, pushing sea levels up several metres.

Stern's conclusion, however, was chilling: to limit the rise to 2 degrees was, in his opinion, too challenging, politically and economically. Instead, he suggested going for a 3-degree target. Yet a 3-degree rise would likely destroy most ecosystems and take global warming beyond the control of human action. It seemed incomprehensible that so few people spoke out forcefully against Stern's target and the death sentence that he was accepting for so many people and species. How could society be so mad as to go for a target that would kill much of the planet's life?

As we started to write our short submission, events in the Arctic intervened to demonstrate dramatically that dangerous climate change is not in the future, but is happening now. Over one northern summer, it became clear that the task was not to weigh up what would be a reasonable rise in temperature; rather, it was to ask: by how much do we need to lower the existing temperature to return our planet to the safe-climate zone? Global warming now demands an emergency response in which we put aside 'business and politics as usual', and devote our collective energy and capacity for innovation to stopping a slide to catastrophe.

Why do business leaders, politicians, community advocates, and sectors of the environment movement fail to grasp fully the momentous problem that we face? How can they not 'get it', when the evidence is now so abundant? We hope that this book will help to identify why real action has so often been blocked, and help to map out a pathway through the barriers.

Many people, including UN secretary-general Ban Ki-Moon, now call the situation that we face a 'climate emergency'; but it could just as easily be called a warming, water, food, or energy emergency. The issues of global warming, water shortages, peak oil, ecosystem destruction, resource depletion, global inequity, and the threat of pandemics intersect and intertwine. Together, these threats and risks constitute a sustainability crisis, or emergency.

In exploring these ideas, our short submission became this unintentional and lengthier book. The story we tell is disturbing and compelling, in equal measure. It poses a choice: to act with great effort now, or to know that it will soon be too late to act effectively. It will be little comfort ten years from now to look back and think ruefully of what we might have done, and of what might have been achieved.

*Climate Code Red* has three interrelated parts.

The first section reviews the scientific evidence and a range of expert insights flowing from the increasing speed of the Arctic sea-ice melt. It also considers recent climate data and analysis about critical subjects such as carbon sinks, biodiversity loss, and climate sensitivity. Drawing on this review, the second section analyses current debates about climate targets, and proposes a set of reduction targets consistent with achieving a safe-climate future. The third section identifies the need for a rapid transition to a sustainable economy. It argues that the task needs to be constructed as a climate emergency—that we cannot continue at the meandering, slow pace dictated by 'business and politics as usual', which today stands in the way of necessary change.

PART ONE
# The Big Melt

'We are all used to talking about these impacts coming in the lifetimes of our children and grandchildren. Now we know that it's us.'
— Professor Martin Parry, co-chairman of the Intergovernmental Panel on Climate Change (IPCC) impacts working group, 18 September 2007.

CHAPTER I

# Losing the Arctic Sea-Ice

In the northern summer of 2007, the area of sea-ice floating on the Arctic Ocean dropped dramatically, reaching the lowest extent that we have seen since records began. This event has profound consequences for climate policy, the role and methods of the Intergovernmental Panel on Climate Change (IPCC), and the assessment of predicted sea-level rises, and it begs the question of whether we have already passed a critical tipping point beyond which human interference with the Earth's climate system becomes very dangerous.

In its February 2007 report on the physical basis of climate science, the IPCC said that Arctic sea-ice was responding sensitively to global warming: 'While changes in winter sea-ice cover are moderate, late summer sea-ice is projected to disappear almost completely towards the end of the twenty-first century.' The IPCC thought it would take a hundred years for the sea-ice to disintegrate; but, even before its report was published, its projections lagged behind the changing physical reality of the Arctic environment.

In a presentation to a Carnegie Institution climate

conference on 1 November 2005, Tore Furevik, professor of oceanography at the Geophysical Institute in Bergen, had already demonstrated that actual Arctic sea-ice retreat in recent years has been greater than had been estimated in any of the 19 Arctic models of the IPCC.

In December 2006, data was presented to an American Geophysical Union (AGU) conference suggesting that the Arctic may be free of all summer sea-ice as early as 2030. According to Marika Holland of the US National Center for Atmospheric Research (NCAR) in Colorado, this would set up a 'positive feedback loop with dramatic implications for the entire Arctic region'. This view was then supported by studies published in March and May 2007 by Holland, along with Mark Serreze and Julienne Stroeve of the US National Snow and Ice Data Centre (NSIDC) at Colorado University.

Soon, events on the ground would outrun even this research. Despite the warnings, experts were shocked at the extent of Arctic ice-sheet loss during the 2007 northern summer. Serreze told the *Guardian* on 4 September: 'It's amazing. It's simply fallen off a cliff and we're still losing ice.'

A feature in the *Washington Post* on 22 October 2007 painted a bleak picture:

> This summer the ice pulled back even more, by an area nearly the size of Alaska. Where explorer Robert Peary just 102 years ago saw 'a great white disk stretching away apparently infinitely' from Ellesmere Island, there is often nothing now but open water. Glaciers race into the sea from the island of Greenland, beginning an inevitable rise in the oceans. Animals are on the move. Polar bears, kings of the Arctic, now search for ice on which to hunt

and bear young. Seals, walrus and fish adapted to the cold are retreating north. New species—salmon, crabs, even crows—are coming from the south. The Inuit, who have lived on the frozen land for millennia, are seeing their houses sink into once-frozen mud, and their hunting trails on the ice are pocked with sinkholes.

On 16 September 2007, the Arctic sea-ice minimum fell to a record low of 4.13 million square kilometres, compared to the previous record low of 5.32 million square kilometres in 2005. This represented a precipitous decline in the ice extent of 22 per cent in two years, an area 'roughly the size of Texas and California combined, or nearly five United Kingdoms,' the NSIDC announced. This compares to the decreasing trend in ice area of 7 per cent *per decade*, between 1979 and 2005.

NSIDC research scientist Walt Meier told the *Independent* on 22 September that the 2007 ice extent was 'the biggest drop from a previous record that we've ever had and it's really quite astounding ... Certainly we've been on a downward trend for the last thirty years or so, but this is really accelerating the trend'.

But it wasn't just the area, or extent, of the sea-ice that was declining rapidly; an even more dramatic story lay hidden beneath. In the early 1960s, the ice was 3.5 metres thick; by the late 1980s, it was down to 2.5 metres; and now, in 2008, large areas are only 1 metre thick. This thinning is accelerating, with half of it having occurred in the past seven years. Taken together, the shrinking ice area and the declining ice thickness mean that the total mass of summer ice has dropped to less than 20 per cent of the volume that it was 30 years ago.

Before the big melt of September 2007, Serreze had

speculated that we may have already reached the tipping point at which there is rapid sea-ice disintegration: 'The big question is whether we are already there or whether the tipping point is still ten or 20 years in the future ... my guts are telling me we may well be there now.' His 'educated guess' was a transition to an ice-free Arctic summer by 2030. His colleague at Colorado Ted Scambos agreed: '2030 is not unreasonable ... I would not rule out 2020, given non-linearity and feedbacks ... I just don't see a happy ending for this.'

These views were supported by Ron Lindsay of the University of Washington: 'Our hypothesis is that we've reached the tipping point. For sea-ice, the positive feedback is that increased summer melt means decreased winter growth and then even more melting the next summer, and so on.'

In 2006, former Australian of the Year, palaeontologist, and climate-change activist Tim Flannery suggested that 'at the trajectory set by the new rate of melt, however, there will be no Arctic icecap in the next five to fifteen years.' By the time September 2007 was over, even these predictions would need to be revised: it was becoming a battle for scientists to keep up with a dramatic climate event that was moving at unimagined speed.

Wieslaw Maslowski of the Naval Postgraduate College in California used US military submarine sonar mapping of the Arctic sea-ice during the many decades of the Cold War in his research, in which he modelled the processes of Arctic sea-ice loss. As early as 2004, he predicted a blue Arctic Ocean free of sea-ice by the summer of 2013, having recognised that the thinning ice was losing volume at a much faster rate than was indicated by satellite-derived surface-area data.

The polar icecap is now floating in water about 3.5

degrees warmer than its historical average. Maslowski found that the sea-ice is being thinned by the northward heat flux of warm summer Pacific and Atlantic waters, not just higher air temperatures. The US National Oceanic and Atmospheric Administration's (NOAA) Arctic report card for 2007 also found that a new wind-circulation pattern is blowing more warm air towards the North Pole than during the previous century; in 2007, winter and spring temperatures were 'all above average throughout the whole Arctic and all at the same time', unlike in previous years.

A team led by Donald Perovich of the Cold Regions Research and Engineering Laboratory in New Hampshire reported in 2007 on research which calculated that, since 1979, the Arctic Ocean has been absorbing sufficient additional heat to melt up to 9300 cubic kilometres of sea-ice, which is equivalent to 3 million square kilometres of ice, 3 metres thick. This is consistent with work by Maslowski and others suggesting that the majority of the ice loss is coming from warmer seas, and not directly from changes in the Arctic climate.

When the ice becomes sufficiently thin, it will be sensitive to a 'kick' from natural climate variations, such as stronger wind and wave-surge action, which will result in rapid loss of the remaining summer ice cover. It could happen any year now. Louis Fortier, scientific director of the Canadian research network ArcticNet, says worst-case scenarios about sea-ice loss are coming true, and that the Arctic Ocean could be ice-free in summertime as soon as 2010. Maslowski told the December 2007 conference of the AGU that 'our projection of 2013 for the removal of ice in summer is not accounting for the last two minima, in 2005 and 2007... So given that fact, you can argue that maybe our projection of 2013 is already too conservative.' NASA climate scientist Jay Zwally told the same conference that, after reviewing recent data, he had concluded that 'the Arctic Ocean could be nearly ice-free at the end of summer by 2012,' while Josefino Comiso of the NASA Goddard Space Centre said: 'I think the tipping point for perennial sea-ice has already passed ... It looks like [it] will continue to decline and there's no hope for it to recover [in the near period].' NASA satellite data shows the remaining Arctic sea-ice is unusually thin, making it more likely to melt in future summers.

While the extent of the sea-ice in winter is about the same as it has been over recent decades, winter ice is becoming increasingly fragile. As the summer extent shrinks, more of the reset winter ice is new, first-year ice. In 2007, only 13 per cent of first-year ice survived the melt season. Because the summer-ice minimum was a record low in 2007, almost two-thirds of the winter ice was first-year ice and, as such, is in a highly vulnerable state. As a result, the northern summer of 2008 is likely to see even more open water.

How many more years will it be before the Arctic is ice-free in summer? The ice retreat is likely to be even bigger each summer, because the refreezing of the ocean surface is starting from a bigger deficit each year, and the amount of older ice is continuing to decrease. Looking at the trends, it is not difficult to see why it may happen in 2009 or 2010, depending on regional weather patterns.

Publicly and privately, many cryosphere climate scientists are shocked and alarmed at these developments. Some Australian climate scientists have expressed similar sentiments, privately acknowledging that the issues of dangerous climate change, caps, and mitigation strategies need urgent review, and that much of the orthodox thinking on these issues is now out of date. What constitutes 'dangerous climate change' is being urgently reviewed.

'The reason so much [of the Arctic ice] went suddenly, is that it is hitting a tipping point that we have been warning about for the past few years,' says the head of NASA's Goddard Institute for Space Studies, James Hansen. He has repeatedly warned that the 'albedo flip' trigger mechanism over large portions of ice sheets could lead to 'eventual non-linear ice sheet disintegration'.

Hansen coined the phrase 'albedo flip' to describe a rapid change, or flip, in the climate that occurs when large areas of ice sheets are lost as a consequence of human-induced warming. In relation to climate, albedo is a measure of the proportion of solar radiation that is reflected (rather than absorbed) when it hits the Earth's surface. White ice reflects most of the radiation (with an albedo of 0.8–0.9), whereas dark surfaces, such as bitumen or dark sea, can have an albedo of less than 0.1 (the Earth's average albedo is 0.3). So when light-reflecting ice sheets are lost and replaced by dark sea, rock surface, or green vegetation, the Earth suddenly absorbs a lot more solar radiation, and the region can heat rapidly. The heating causes more regional ice-sheet disintegration, in a classic example of 'positive feedback'. This is now occurring in the Arctic. The eventual consequence is higher global temperatures, which we estimate will increase by around 0.3 degrees.

There are other factors that impact on the albedo effect. As ice is lost and regional temperatures increase, the atmosphere holds more water vapour and the cloud-cover increases, and this 'negative feedback' cancels out some of the warming effect from the loss of reflective sea-ice. General climate models do show that the cloud feedback can damp, or delay, the climate system's response, but they also show that surface albedo changes far outweigh the influence of cloud changes. There are no general climate model studies which specifically address the question of the global warming that is likely to occur as a consequence of the total loss of the Arctic sea-ice and its albedo effect, but a finding of a 0.01 degree warming for every 1 per cent of ice-sheet loss indicates that 0.3 degrees would be the eventual global warming that would result from total Arctic sea-ice loss. A higher figure may be more consistent with recent evidence, including typical temperature flips in glacial–interglacial cycles. On the other hand, there is a small chance that later this century an abrupt change in the North Atlantic thermohaline circulation (the 'Gulf Stream'), caused by global warming, could generate a strong cooling effect along the Norwegian coast, which would lead to the re-establishment of year-round Arctic sea-ice cover.

But there is much support for Hansen's albedo-flip predictions, including from Pål Prestrud of Oslo's Center for International Climate and Environmental Research, who says: 'We are reaching a tipping point, or are past it, for the ice. This is a strong indication that there is an amplifying mechanism here.'

The 2007 summer Arctic Ocean surface temperatures are estimated to have been much warmer than in previous years, by up to 5 degrees, largely as a result of solar heating of the upper ocean, which was the result of less cloud cover increasing the albedo feedback. 'That feedback is the key to why the models predict that the Arctic warming is going to be faster,' Zwally told the 2007 AGU meeting. 'It's getting even worse than the models predicted.'

There are also questions about the quantity of methane and carbon dioxide that will be released as a consequence of rapid regional warming, as areas of tundra permafrost in the polar north defrost. One initial estimate is that carbon dioxide released from permafrost may contribute an extra 0.7 of a degree over the next 100 years, based on predicted global warming of 2 degrees. But there is uncertainty about these figures, because they depend on how much extra positive feedback is included, for example, from water vapour; how much permafrost is thawed; and how much carbon is released. This figure is a conservative estimate, because it does not include the warming effect of methane release.

The Arctic summer of 2007 demands that we look in detail at its consequences. Sea-ice loss and Arctic warming are affecting the permafrost in Siberia, Alaska, and other regions; they are triggering caribou decline in Canada; and they are 'shrubifying' the tundra. 'What happens in the Arctic actually does not stay in the Arctic,' says oceanographer Richard Spinrad, who is deputy chief of the NOAA. In this case, one of the most significant knock-on impacts is going to be on the Greenland ice sheet.

CHAPTER 2

# Greenland's Fate

Covering four-fifths of the landmass of Greenland is an ice sheet almost 2400 kilometres long, more than 2 kilometres thick, and 1100 kilometres across at its widest point. It contains more than 2.8 million cubic kilometres of ice. If it were to melt entirely, it would raise sea levels by more than 7 metres. The question is, will it melt—and how quickly?

The 2001 IPCC reports suggested that neither the Greenland nor the Antarctic ice sheets would lose significant mass by 2100. By the final IPCC report for 2007, this degree of certainty was evaporating, with a view that 'uncertainties ... in the full effects of changes in ice sheet flow' contributed to an unwillingness to put an upper bound on sea-level rises this century. The IPPC noted that 'partial loss of ice sheets on polar land could imply metres of sea level rise ... Such changes are projected to occur over millennial time scales, but more rapid sea level rise on century time scales cannot be excluded'. It suggested that the Greenland ice sheet would be virtually eliminated, and would result in a sea-level rise of 7 metres, if 'global average warming were sustained for

millennia in excess of 1.9 to 4.6 degrees'. In other words, the IPCC felt that if the disintegration of the Greenland ice sheet were to be triggered, the process of loss would take a thousand years or more to conclude.

This understanding was soon superseded when it was found that widening cracks in the massive ice sheet were producing an unexpected effect: as surface-melt water ran across the ice, it formed streams that widened into a torrent of water which was pouring through the cracks and carving a wider cavity as it rushed, like a giant waterfall, into an icy void below.

An already famous photograph by Roger Braithwaite of the University of Manchester (see previous page) shows the diminutive figures of his fellow scientists, dwarfed by a rapidly flowing gully of water several metres wide that gathered speed and then disappeared into one of these cavities, winding its way down many hundreds of metres through a honeycomb that the cracking and the water flow had opened up. Reaching the base of the ice sheet, the water acted to lubricate the movement of the ice sheet over the rocky bottom.

This new 'wet' process helps to explain why Greenland is losing ice mass more quickly than predicted. Two studies published in 2006 (one led by Eric Rignot of NASA's Jet Propulsion Laboratory in Pasadena, and another from the University of Texas) found that the Greenland ice cap 'may be melting three times faster than indicated by previous measurements', and that 'the mass loss is increasing with time'. Rignot told *New Scientist*: 'These results absolutely floored us … The glaciers are sending us a signal. Greenland is probably going to contribute more and faster to sea-level rise than predicted by current models.'

Greenland experienced more days of melting snow in 2006 than the island had averaged over recent decades. According to NASA researcher Marco Tedesco, the area experiencing at least one day of melting has been increasing since 1992 at a rate of 35,000 square kilometres per year, and the melt extent for 2007 was the largest recorded since satellite measurements began, in 1979. Thomas Mote of the Climatology Research Laboratory at the University of Georgia found the summertime melt in 2007 to be the most extreme so far: 60 per cent greater than the previous highest rate, in 1998. The edges of the ice sheet are melting up to ten times more rapidly than earlier research had indicated, and the ice-sheet height is falling by up to 10 metres a year.

The University of Colorado's Konrad Steffen says that air temperatures on the Greenland ice sheet have increased by 4 degrees since 1991, and that the trend toward increases in the total area of bare ice that is subject to at least one day's melting per year is unmistakeable, at 13 per cent per year. Steffen says the ice-loss trend in Greenland is similar to the trend of Arctic sea-ice in recent decades, suggesting the rate of loss may be increasing exponentially.

Robert Corell, the chair of the IPCC's Arctic Climate Impact Assessment, says of Greenland: 'Nobody knows now how quickly it will melt ... This is all unprecedented in the science ... Until recently, we didn't believe it possible, for instance, for water to permeate a glacier all the way to the bottom. But that's what's happening. As the water pools, it opens more areas of ice to melting.' As the ice sheet cracks into huge pieces that are several cubic kilometres in size, it scrapes across the bedrock and triggers earthquake-like tremors. And there is an acceleration of the speed at which

Greenland glaciers are moving into the sea; some glacier velocities have more than doubled. The Jacobshavn Glacier, a major Greenland outlet glacier that drains roughly 8 per cent of the ice sheet, has sped up nearly twofold in the last decade. Another indication of Greenland's shrinking ice cap is evidence that its landmass is rising by up to 4 centimetres per year, a buoyancy produced by carrying less weight of ice.

At some point, the trajectory of ice disintegration may pass a tipping point for the loss of, at least, a substantial portion of Greenland's ice sheet.

Global warming has so far been greatest in the high latitudes of the northern hemisphere, particularly in the sub-Arctic forests of Siberia and North America. And because Arctic temperatures rise more than twice as much as the global average, a global warming of 2 degrees is likely to result in average annual warming over the Arctic of 3.2–6.6 degrees. With Greenland's critical-melt threshold estimated to be a regional warming of 2.7 degrees, this point will be triggered by a global rise of just over 1 degree — a rise that is now considered impossible to avoid.

Tipping points may be looming, and we may not be aware of it. Is Greenland such a case? Before the full extent of the dramatic Arctic sea-ice loss for 2007 was known, Tim Lenton of the University of East Anglia told a Cambridge conference that 'we are close to being committed to a collapse of the Greenland ice sheet'. In recognition of the limits of climate science models, London School of Economics statistician Lenny Smith told the same conference that 'we need to drop the pretence that [the models] are nearly perfect'. He said that there were too many 'unknown unknowns', and that 'we need to be more open about our uncertainties'.

According to a 2006 report in *New Scientist* magazine, rising Arctic regional temperatures are already at 'the threshold beyond which glaciologists think the [Greenland] ice sheet may be doomed'.

But the issue is disputed, because the orthodox climate and ice-loss models for Greenland do not include the processes that result in meltwater penetrating crevasses and lubricating the glaciers' flow. So how can we know at what pace the loss of ice volume is likely to proceed? James Hansen says ice-sheet disintegration 'starts slowly, but multiple positive feedbacks can lead to rapid non-linear collapse'. He says that there is the potential for us to lose control, because we 'cannot tie a rope around a collapsing ice sheet'.

Hansen identifies a scientific reticence that, in at least some cases, 'hinders communication with the public about dangers of global warming'. He says, 'Scientific reticence may be a consequence of the scientific method. Success in science depends on objective scepticism. Caution, if not reticence, has its merits; however, in a case such as ice sheet instability and sea-level rise, there is a danger in excessive caution. We may rue reticence, if it serves to lock in future disasters.'

But there are useful sources, other than models, for thinking about the likely future rate of loss of the Greenland ice sheet, including expert assessment and paleoclimatology (the study of ancient climates). In response to the concerns expressed by Corell, Hansen, and many others about the IPCC process and the currency of some of its 2007 report, Michael Oppenheimer and his colleagues from Princeton University have proposed that the inputs into IPCC reports be broadened. They say that the observational data, past climate record, and theoretical evidence of poorly understood phenomena should be given 'a comparable weight with evidence from numerical modelling'. Oppenheimer notes that, in areas in which modelling evidence is sparse or lacking, the IPCC sometimes provides no uncertainty estimate at all; and, in other areas, the use of models with similar structures leads to an artificially high confidence in projections that is not warranted. He also calls on the IPCC to fully include judgements from experts.

In the absence of reliable, computer-based models of the workings of the Greenland ice sheet, what do the experts think? Hansen asks:

> Could the Greenland ice sheet survive if the Arctic were ice-free in summer and fall? It has been argued that not only is ice-sheet survival unlikely, but its disintegration would be a wet process that can proceed rapidly. Thus an ice-free Arctic Ocean, because it may hasten [the] melting of Greenland, may have implications for global sea level, as well as the regional environment, making Arctic climate change centrally relevant to [the] definition of dangerous human interference.

Arctic climate researchers will say, off the record, that this is not an unreasonable view; on the record, they will say that there are no verifiable models that produce this result. These statements are not in contradiction.

As for the paleoclimate record, global average temperatures have now surpassed those that thawed much of Greenland's ice cap some 130,000 years ago, when the planet experienced a warm interlude from continent-covering glaciers, and seas were 5–6 metres higher than today. Global warming appears to be pushing vast reservoirs of ice on Greenland and Antarctica toward a significant long-term meltdown. The world may have as little as a decade to take the steps to avoid this scenario, says Hansen.

Following the extraordinary Arctic summer of 2007, along with the arrival of new data, a number of leading climate scientists have spoken out, most notably at the December 2007 AGU meeting. There, Hansen said that, based on Greenland melt data, the Earth has hit one of its tipping points, and that the level of carbon dioxide in the atmosphere is now enough to cause Arctic sea-ice cover and massive ice sheets, such as those on Greenland, eventually to melt away.

In summary, in the polar north it is reasonable to expect rapid loss of the Arctic sea-ice, with a significant impact on regional temperatures. It is also reasonable to expect, as a consequence, an acceleration of the rate of loss of Greenland ice, which may already be past its disintegration tipping point for a large part of the ice sheet—a situation that had previously been predicted to occur a long way into the future.

CHAPTER 3

# Trouble in the Antarctic

Big changes are also underway at the other end of the world, in the Antarctic, where most of the world's ice sits on the fifth-largest continent. The majority of Antarctic ice is contained in the East Antarctic ice sheet—the biggest slab of ice on Earth, which has been in place for some 20 million years and which, if fully melted, would raise sea levels by more than 60 metres.

Considered more vulnerable is the smaller West Antarctic ice sheet, which contains one-tenth of the total Antarctic ice volume. If it disintegrated, it would raise sea levels by around 5 metres, a similar amount to what we would see with a total loss of the Greenland ice sheet.

While it was generally anticipated that the West Antarctic sheet would be more stable than Greenland at a 1–2 degree rise, recent research demonstrates that the southern ice shelf reacts far more sensitively to warming temperatures than scientists had previously believed. Ice-core data from the Antarctic Geological Drilling joint project (being conducted by Germany, Italy, New Zealand, and the United States) shows that 'massive melting' must have occurred in the Antarctic

28

three million years ago, during the Miocene–Pliocene period, when the average global temperature in the oceans increased by only 2–3 degrees above the present temperature. Geologist Lothar Viereck-Götte called the results 'horrifying', and suggested that 'the ice caps are substantially more mobile and sensitive than we had assumed'.

The heating effect caused by climate change is greatest at the poles, and the air over the West Antarctic peninsula has warmed nearly 6 degrees since 1950. At the same time, according to a report in the *Washington Post* on 22 October 2007, a warming sea is melting the ice-cap edges, and beech trees and grass are taking root on the ice fringes.

Another warning sign was the rapid collapse in March 2002 of the 200-metre-thick Larsen B ice shelf, which had been stable for at least twelve thousand years, and which was the main outlet for glaciers draining from West Antarctica. An ice shelf is a floating sheet, or platform, of ice. Largely submerged, and up to a kilometre thick, the shelf abuts the land and is formed when glaciers or land-based ice flows into the sea. Generally, an ice shelf will lose volume by calving icebergs, but these are also subject to rapid disintegration events. Larsen B, weakened by water-filled cracks where its shelf attached to the Antarctic Peninsula, gave way in a matter of days, releasing five hundred billion tonnes of ice into the ocean.

Neil Glasser of Aberystwyth University and Ted Scambos from the NSIDC found that as glacier flow had begun to increase during the 1990s, the ice shelf had become stressed. The warming of deep Southern Ocean currents (which increasingly reach the Antarctic coastline) had also led to some thinning of the shelf, making it more prone to breaking apart. Scambos concludes that 'the unusually warm summer of 2002, part of a multi-decade trend of warming [that is] clearly tied to climate change, was the final straw'.

Looking at the overall pace of events, Scambos says: 'We thought the southern hemisphere climate is inherently more stable, [but] all of the time scales seem to be shortened now. These things can happen fairly quickly. A decade or two of warming is all you need to really change the mass balance ... Things are on more of a hair trigger than we thought.'

Much of the West Antarctic ice sheet sits on bedrock that is below sea level, buttressed on two sides by mountains, but held in place on the other two sides by the Ronne and Ross ice shelves; so, if the ice shelves that buttress the ice sheet disintegrate, sea water breeching the base of the ice sheet will hasten the rate of disintegration.

In 1968, the Ohio State University glaciologist John Mercer warned, in the journal of the International Association of Scientific Hydrology, that the collapse of ice shelves along the Antarctic Peninsula could herald the loss of the ice sheet in West Antarctica. A decade later, in 1978, his views received a wider audience in *Nature*, where he wrote: 'I contend that a major disaster—a rapid deglaciation of West Antarctica—may be in progress ... within about 50 years.' Mercer said that warming 'above a critical level would remove all ice shelves, and consequently all ice grounded below sea level, resulting in the deglaciation of most of West Antarctica'. Such disintegration, once under way, would 'probably be rapid, perhaps catastrophically so', with most of the ice sheet lost in a century. Credited with coining the phrase 'the greenhouse effect' in the early 1960s, Mercer's Antarctic prognosis was widely ignored and disparaged at the time, but this has changed.

(James Hansen says it was not clear at the time whether

Mercer or his many critics were correct, but those who labelled Mercer an alarmist were considered more authoritative and better able to get funding. Hansen believes funding constraints can inhibit scientific criticisms of the status quo. As he wrote in *New Scientist* on 28 July 2007: 'I believe there is pressure on scientists to be conservative.' Hansen is responsible for coining the term 'The John Mercer Effect', meaning to play down your findings for fear of losing access to funding or of being considered alarmist.)

Another vulnerable place on the West Antarctic ice sheet is Pine Island Bay, where two large glaciers, Pine Island and Thwaites, drain about 40 per cent of the ice sheet into the sea. The glaciers are responding to rapid melting of their ice shelves and their rate of flow has doubled, whilst the rate of mass loss of ice from their catchment has now tripled. NASA glaciologist Eric Rignot has studied the Pine Island glacier, and his work has led climate writer Fred Pearce to conclude that 'the glacier is primed for runaway destruction'. Pearce also notes the work of Terry Hughes of the University of Maine, who says that the collapse of the Pine Island and Thwaites glaciers—already the biggest causes of global sea-level rises—could destabilise the whole of the West Antarctic ice sheet. Pearce is also swayed by geologist Richard Alley, who says there is 'a possibility that the West Antarctic ice sheet could collapse and raise sea levels by 6 yards [5.5 metres]', this century.

Hansen and fellow NASA Goddard Institute for Space Studies researcher Makiko Sato agree:

The gravest threat we foresee starts with surface melt on West Antarctica, and interaction among positive feedbacks leading to catastrophic ice loss. Warming in

West Antarctica in recent decades has been limited by effects of stratospheric ozone depletion. However, climate projections find surface warming in West Antarctica and warming of nearby ocean at depths that may attack buttressing ice shelves. Loss of ice shelves allows more rapid discharge from ice streams, in turn a lowering and warming of the ice sheet surface, and increased surface melt. Rising sea level helps unhinge the ice from pinning points ... Attention has focused on Greenland, but the most recent gravity data indicate comparable mass loss from West Antarctica. We find it implausible that BAU ['business-as-usual'] scenarios, with climate forcing and global warming exceeding those of the Pliocene, would permit a West Antarctic ice sheet of present size to survive even for a century.

Even in East Antarctica, where total ice loss would produce a sea-level rise of 60 metres, mass loss near the coast is greater than the mass increase inland (mass increase inland is caused by the extra snowfall generated from warming-induced increases in air humidity).

While the inland of East Antarctica has cooled during the last 20 years, the coast has become warmer, with melting occurring 900 kilometres from the coast and in the Transantarctic Mountains, which rise up to an altitude of 2 kilometres.

Research published in January 2008 by Rignot and six of his colleagues shows that ice loss in Antarctica has increased by 75 per cent in the last ten years due to a speed-up in the flow of its glaciers, so that the ice loss there is now nearly as great as that observed in Greenland.

# CHAPTER 4

# A Rising Tide

Many climate scientists received the 2007 IPPC report's suggestion of a sea-level rise of 18–59 centimetres by 2100 with dismay—not because it was too high, but because it seriously underestimated the problem. Before the report was released, satellite data showed sea levels had risen by an average of 3.3 millimetres per year between 1993 and 2006; the 2001 IPCC report, by contrast, projected a best-estimate rise of less than 2 millimetres per year.

Now three researchers from Taiwan's National Central University have calculated that, in the last half-century, water run-off impounded on land (principally in dams) would have raised sea levels by another 3 centimetres, or an average of 0.55 millimetres per year, if it had been allowed to reach the sea. This means that the contribution to sea-level rises from ice-sheet disintegration has been underestimated, which leads to the conclusion that the runoff from melting mountain glaciers, including the Himalayas, is much greater than previously thought.

In late 2006, work by Stefan Rahmstorf of the Potsdam

Institute for Climate Impact Research (PICIR) concluded that previous estimates of how much the world's sea level will rise as a result of global warming might have been seriously underestimated. Rahmstorf, an ocean physicist, said that the data now available 'raises concerns that the climate system, in particular sea level, may be responding more quickly than climate models indicate'. Even before the IPCC's first report for 2007 was released in February, Robert Corell said that any prediction of a sea-level rise by 2100 of less than a metre would 'not be a fair reflection of what we know'.

The final IPCC 2007 report, released on 16 November, drew together individual reports published earlier in the year. It contained the following qualification: 'Because understanding of some important effects driving sea-level rise is too limited, this report does not assess the likelihood, nor provide a best estimate or an upper bound for sea level rise.' It added that the official projected sea-level rise of 18–59 centimetres this century did 'not include uncertainties in climate-carbon cycle feedbacks nor the full effects of changes in ice sheet flow, therefore the upper values of the ranges are not to be considered upper bounds for sea level rise'. This begs the question: why were the official projections included at all if, in this innovative turn of phrase, 'the upper values of the ranges are not to be considered upper bounds'?

So, how much will sea levels rise this century? At what rate will the Greenland and West Antarctic ice sheets disintegrate, and what influence will the earlier-than-anticipated loss of the Arctic sea-ice have on Greenland's rate of loss? These questions have caused turmoil in scientific circles, because it is generally acknowledged that the sea-level rise will be a good deal higher than the early-2007 IPCC report suggests.

The IPCC's projection of a sea-level rise of well under a metre by 2100 was based on models that IPCC critics, including Michael Oppenheimer of Princeton, say '[do not] include the potential for increasing contributions from rapid dynamic processes in the Greenland and West Antarctic ice sheets, which have already had a significant effect on sea level over the past 15 years and could eventually raise sea level by many meters'. They say that, lacking such processes, 'models cannot fully explain observations of recent sea level rise, and accordingly, projections based on such models may seriously understate potential future increases'.

Writing in the *Guardian* on 7 September 2007, environment correspondent Paul Brown reported that 'scientists monitoring [Arctic] events this summer say the acceleration could be catastrophic in terms of sea-level rise and make predictions this February [2007] by the [IPCC] far too low'. This topic is now the subject of urgent collaborative work between a number of agencies and research centres. There is also new concern about the contribution to sea-level rises that may be made by the West Antarctic ice sheet.

James Hansen and his collaborators have taken a leading role in this discussion, in a number of recent peer-reviewed papers. Their essential argument, based on the paleoclimate record, is that the sea-level rise is likely to be about 5 metres this century if emissions continue down the 'business as usual' trajectory.

Here is the core of Hansen's expert opinion:

I find it almost inconceivable that 'business as usual' climate change will not result in a rise in sea level measured in meters within a century ... Because while the growth of great ice sheets takes millennia, the disintegration of ice sheets is a wet process that can proceed rapidly ...

[T]he primary issue is whether global warming will reach a level such that ice sheets begin to disintegrate in a rapid, non-linear fashion on West Antarctica, Greenland or both. Once well under way, such a collapse might be impossible to stop, because there are multiple positive feedbacks. In that event, a sea-level rise of several meters at least would be expected.

As an example, let us say that ice sheet melting adds 1 centimeter to sea level for the decade 2005 to 2015, and that this doubles each decade until the West Antarctic ice sheet is largely depleted. This would yield a rise in sea level of more than 5 meters by 2095.

Of course, I cannot prove that my choice of a 10-year doubling time is accurate but I'd bet $1000 to a doughnut that it provides a far better estimate of the ice sheet's contribution to sea-level rise than a linear response. In my opinion, if the world warms by 2°C to 3°C, such massive sea-level rise is inevitable, and a substantial fraction of the rise would occur within a century. 'Business as usual' global warming would almost surely send the planet beyond a tipping point, guaranteeing a disastrous degree of sea-level rise.

Although some ice sheet experts believe that the ice sheets are more stable, I believe that their view is partly based on the faulty assumption that the Earth has been as much as 2 °C warmer in previous interglacial periods, when the sea level was at most a few meters higher than at present. There is strong evidence that the Earth now is within 1°C of its highest temperature in the past million years. Oxygen isotopes in the deep-ocean fossil plankton known as foraminifera reveal that the Earth was last 2°C to 3°C warmer around 3 million years ago, with carbon dioxide levels of perhaps 350 to 450 parts per million. It was a dramatically different planet then, with no Arctic sea-ice in the warm seasons and sea level about 25 meters higher, give or take 10 meters.

There is not a sufficiently widespread appreciation of the implications of putting back into the air a large fraction of the carbon stored in the ground over epochs of geologic time. The climate forcing caused by these greenhouse gases would dwarf the climate forcing for any time in the past several hundred thousand years — the period for which accurate records of atmospheric composition are available from ice cores.

Models based on the 'business as usual' scenarios of the Intergovernmental Panel on Climate Change (IPCC) predict a global warming of at least 3°C by the end of this century. What many people do not realise is that these models generally include only fast feedback processes: changes in clouds, water vapour and aerosols. Actual global warming would be greater as slow feedbacks come into play: increased vegetation at high latitudes, ice sheet shrinkage and further greenhouse gas emissions from the land and sea in response to global warming.

The IPCC's latest projection for sea-level rise this century is 18 to 59 centimeters. Though it explicitly notes that it was unable to include possible dynamical responses of the ice sheets in its calculations, the provision of such specific numbers encourages a predictable public belief that the projected sea level change is moderate, and indeed smaller than in the previous IPCC report. There have been numerous media reports of 'reduced' predictions of sea-level rise, and commentators have denigrated suggestions that 'business as usual' emissions may cause a sea-level rise measured in meters. However, if these IPCC numbers are taken as predictions of actual sea-level rise, as they have been by the public, they imply that the ice sheets can miraculously survive a 'business as usual' climate forcing assault for a millennium or longer.

There are glaciologists who anticipate such long response times, because their ice sheet models have been designed to match past climate changes. However, work by my group shows that the typical 6000-year timescale for ice sheet disintegration in the past reflects the gradual changes in Earth's orbit that drove climate changes at the time, rather than any inherent limit for how long it takes ice sheets to disintegrate. Indeed, the paleoclimate record contains numerous examples of ice sheets yielding sea-level rises of several meters per century when forcings were smaller than that of the 'business as usual' scenario. For example, about 14,000 years ago, sea level rose approximately 20 meters in 400 years, or about 1 meter every 20 years.

There is growing evidence that the global warming already under way could bring a comparably rapid rise in sea level. The process begins with human-made greenhouse gases, which cause the atmosphere to be more opaque to infrared radiation, thus decreasing radiation of heat to space. As a result, the Earth is gaining more heat than it is losing: currently 0.5 to 1 watts per square meter. This planetary energy imbalance is sufficient to melt ice corresponding to 1 meter of sea-level rise per decade, if the extra energy were used entirely for that purpose—and the energy imbalance could double if emissions keep growing.

So where is the extra energy going? A small part of it is warming the atmosphere and thus contributing to one key feedback on the ice sheets: the 'albedo flip' that occurs when snow and ice begin to melt. Snow-covered ice reflects back to space most of the sunlight striking it, but as warming air causes melting on the surface, the darker ice absorbs much more solar energy. This increases the planetary energy imbalance and can lead to more melting. Most of the resulting meltwater burrows through the ice sheet, lubricating its base and speeding up the discharge of icebergs to the ocean.

The area with summer melt on Greenland has increased from around 450,000 square kilometers when satellite observations began in 1979 to more than 600,000 square kilometers in 2002. Seismometers around the world have detected an increasing number of earthquakes on Greenland near the outlets of major ice streams. The earthquakes are an indication that large pieces of the ice sheet lurch forward and then grind to a halt because of friction with the ground. The number of these 'ice quakes' doubled between 1993 and the late 1990s, and it has since doubled again. It is not yet clear whether the quake number is proportional to ice loss, but the rapid increase is cause for concern about the long-term stability of the ice sheet.

Additional global warming of 2°C to 3°C is expected to cause local warming of about 5°C over Greenland. This would spread summer melt over practically the entire ice sheet and considerably lengthen the melt season. In my opinion it is inconceivable that the ice sheet could withstand such increased meltwater for long before starting to disintegrate rapidly, but it is very difficult to predict when such a period of large, rapid change would begin.

Summer melt on West Antarctica has received less attention than on Greenland, but it is more important. The West Antarctic ice sheet, which rests on bedrock far below sea level, is more vulnerable as it is being attacked from below by warming ocean water, as well as from above by a warming atmosphere. Satellite observations reveal increasing areas of summer melt on the West Antarctic ice sheet, and also a longer melt season.[*]

[from 'Huge sea level rises are coming—unless we act now', *New Scientist*, 28 July 2007]

Hansen's argument has been cited at length here because he is one of the world's most influential and outspoken climate scientists. He has provided a compelling critique of the limitations of the IPCC models, and has helped to evolve a new understanding of the mechanics of rapid ice-sheet disintegration. He played a significant role in public debate in the US about global warming, and twice testified before the Congress on climate change: the first occasion, in 1987, caused a minor controversy when Hansen insisted that his group of climate modellers could 'confidently state that major greenhouse climate changes are a certainty', and that 'the global warming predicted in the next 20 years will make the Earth warmer than it has been in the past 100,000 years'.

Hansen has also endured having funds for the NASA

---

[*] All warmings in this quoted material are relative to the temperature in 2000.

Goddard Institute for Space Studies slashed by the Bush administration, because he refused to stop his public advocacy on climate action and policy. Now he has staked his professional reputation on the issue of the speed of sea-level rises, with his preparedness, in his own words, to bet '$1000 to a doughnut' that his view is closer to the mark than the view of the IPCC.

The incongruity of the IPCC's sea-level projection for 2100 can be seen in the following figure, which illustrates mean global temperature and sea level (relative to today) at different times in Earth's history, alongside the IPCC projection for 2100 (the outline circle). For the longer term, the paleoclimate data suggests a much higher sea-level rise at equilibrium than that projected by the IPCC for this century.

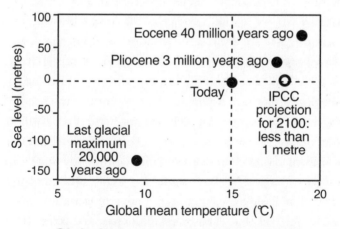

## Global temperature and sea level

Mean global temperature and sea level (relative to today's) at different times in Earth's history, compared with the IPCC projection for the year 2100. Courtesy David Archer.

Now, with new Arctic data, Hansen has firmed his outlook. He told the December 2007 AGU meeting that there is 'already enough carbon in Earth's atmosphere to ensure that sea levels will rise several feet in coming decades', because the world, including Greenland, passed the tipping point for major ice-sheet loss decades ago.

But long before the Greenland or West Antarctic ice sheets fully disintegrate, even the loss of 20 per cent of Greenland's ice volume would be catastrophic. In his 2006 report to the UK government, Sir Nicholas Stern noted that:

[C]urrently, more than 200 million people live in coastal floodplains around the world, with two million square kilometres of land and one trillion dollars worth of assets less than one metre elevation above current sea level. One-quarter of Bangladesh's population (~35 million people) lives within the coastal floodplain. Many of the world's major cities (22 of the top 50) are at risk of flooding from coastal surges, including Tokyo, Shanghai, Hong Kong, Mumbai, Kolkata, Karachi, Buenos Aires, St Petersburg, New York, Miami and London. In almost every case, the city relies on costly flood defences for protection. Even if protected, these cities would lie below sea level with a residual risk of flooding like New Orleans today. The homes of tens of millions more people are likely to be affected by flooding from coastal storm surges with rising sea levels. People in South and East Asia will be most vulnerable, along with those living on the coast of Africa and on small islands.

Underground water is the largest reserve of fresh water

on the planet, and more than two billion people depend on it. Long before rising seas inundate the land, aquifers will be contaminated. The 2006 conference of the International Association of Hydrogeologists heard that rising sea levels would lead to salt-water inundation of the aquifers used by cities such as Shanghai, Manila, Jakarta, Bangkok, Kolkata, Mumbai, Karachi, Lagos, Buenos Aires, and Lima. Fred Pearce described this problem in *New Scientist* on 16 April 2006:

> The water supplies of dozens of major cities around the world are at risk from a previously ignored aspect of global warming. Within the next few decades rising sea levels will pollute underground water reserves with salt … Long before the rising tides flood coastal cities, salt water will invade the porous rocks that hold fresh water … The problem will be compounded by sinking water tables due to low rainfall, also caused by climate change, and rising water usage by the world's growing and increasingly urbanised population.

In the low-lying eight-island Pacific nation of Tuvalu, contamination of water for domestic and agricultural use is already serious, and it is a growing concern in other regional states, including Kiribati and the Marshall Islands.

While large sea-level-rise figures may seem abstract, a rise of 1 metre will have a devastating impact on densely populated river deltas in the developing world, as homes and agricultural land are lost and damaged by storm surges. In industrialised regions, small rises will have severe impacts on coastal infrastructure: loss of beaches, ports, and shipping facilities; flooding of transport links; inundation of underground facilities, including sewers, water, electricity transmission, and communications infrastructure; as well as the loss of industrial and domestic buildings.

The degree to which this rise would physically impinge on these coastal areas is hard to imagine, but one can get a powerful visual sense of it by using Google Maps with a sea-level rise overlay (for example, lood.firetree.net), which is instructive for understanding the lessons from the Arctic summer of 2007, and from West Antarctica.

CHAPTER 5

# The Quickening Pace

How sensitive is the Earth to changes in greenhouse gases and other influences that shape or 'force' the climate? In 1977, meteorologist Jule Charney developed what has become the yardstick for comparing estimates of 'climate sensitivity' (which is the warming estimated to occur if carbon dioxide levels were doubled from pre-industrial levels).

Climate-sensitivity research has produced quite divergent results, particularly earlier on, but a warming of 3 degrees is now the widely accepted estimate, known as the 'Charney 3°C'. The main point is that, the higher the figure, the more trouble we are going to be in.

The 2007 IPCC report uses Equilibrium Climate Sensitivity (ECS) models to conclude that the warming is 'likely to be in the range of 2 to 4.5 degrees with a best estimate of about 3 degrees, and is very unlikely to be less than 1.5 degrees. Values substantially higher than 4.5 degrees cannot be excluded, but agreement of models with observations is not as good for those values'. Nevertheless, a number of studies have found larger possible ranges, including ranges of 1–10 degrees and

1.4–7.7 degrees. One study even suggested that there is a 54 per cent likelihood that climate sensitivity lies outside the IPCC range.

In 2006, Barrie Pittock, a former senior climate scientist for Australia's Commonwealth Scientific and Industrial Research Organisation (CSIRO), suggested that the 'dated IPCC view might underestimate the upper end of the range of possibilities', adding that there is 'a much higher probability of warmings by 2100 exceeding the mid-level [climate sensitivity] estimate of 3 degrees'. He wrote in the journal *Ecos* that recent data suggests critical levels of global warming may occur at even lower greenhouse-gas concentrations than were considered to be the danger level in the 2001 IPCC report, and identifies at least eight areas of concern. These include the lessening of global dimming, increased permafrost melting, release of soil carbon, Arctic sea-ice retreat, circulation change in mid to high latitudes, rapid changes in Greenland and Antarctica, increasing intensity of tropical cyclones, and a slowing of the Gulf Stream.

A doubling of greenhouse gases would take the pre-industrial carbon dioxide level of 280 parts per million to 560 parts per million. Given that carbon dioxide levels in the atmosphere are now at 387 parts per million, we are already one-third of the way to that doubling. And if we take into account the warming of 0.8 degrees that we've experienced so far, and add the delayed warming (still in the pipeline) of 0.6 degrees, along with the 0.3-degree effect of the Arctic 'albedo flip', current climate-energy imbalances alone will, eventually, warm the planet by around 1.7 degrees. It seems likely, then, that the global temperature-increase which will occur at 560 parts per million of carbon dioxide will be a lot more than the mid-range climate-sensitivity estimate of 3 degrees—especially if major ice sheets and rainforests are lost, and carbon sinks (such as the oceans and the soils, which absorb and release carbon as parts of the natural cycle) continue to weaken along the current trends.

The 3-degree sensitivity estimate only takes into account 'fast' feedbacks that come into play quickly. Water vapour, for example, is a greenhouse gas, and the air holds more water vapour as the temperature increases, which produces a fast feedback. Other fast feedbacks include changes in cloud cover, snow cover, and sea-ice extent.

The problem is that ECS models omit 'slow' feedbacks, such as ice sheet growth and decay, change in vegetation cover, permafrost melting and methane release, and carbon-cycle feedbacks, all of which amplify climate changes on time scales of decades to centuries.

Hansen and Sato argue that, while the 'Charney 3°C' is reasonable in the short run, the total long-term climate sensitivity for fast and slow feedbacks is likely to be 'about 6 degrees for doubled carbon dioxide', based on the paleoclimate data. In terms of the time scale of the last three decades, they say:

> [T]he Charney sensitivity is a good approximation, as little contribution from slow feedbacks would be expected. Thus climate models with 3-degree sensitivity for doubled carbon dioxide, incorporating only the fast feedbacks, are able to achieve good agreement with observed warming of the past century. We suggest, however, [that] these models provide only a lower limit on the expected warming on century time scales due to the assumed forcings. The

real world will be aiming in the longer run at a warming corresponding to the higher climate sensitivity [of 6 degrees].

Hansen and Sato say that 'slower' feedbacks (including the pole-ward movement of forests, shrinking and loss of ice sheets, and release of methane from melting tundra) are likely to be significant on decade-to-century timescales, as we are now starting to witness. Those slower feedbacks mean that coming to an understanding of what would constitute 'dangerous' climate change becomes more urgent, as does finding a path to avoid it.

Paleoclimate data identifies the impact that these missing slow feedbacks have in pushing temperatures higher than expected. New research matching greenhouse-gas levels with the Earth's temperature over the last 450,000 years has established the climate sensitivity with slow feedbacks to be 6 degrees. Fifty-five million years ago, in the Arctic, temperatures were 11 degrees warmer than the ECS models would predict—which also suggests that other feedback mechanisms were at work.

A paper published in the June 2005 issue of *Nature* supports the theory of even higher climate sensitivity. It describes research led by Meinrat Andreae of the Max Planck Institute for Chemistry in Germany, which used climate models and various aerosol-cooling assumptions to find the 'best fit' for the data involved in a climate sensitivity in excess of 6 degrees. By studying the planet's climate history over the last 50 years and fitting it to various climate-model options, they concluded that the effects of airborne particle pollution (or aerosols: soot and exhausts from burning fossil fuels, industrial pollution, and dust storms) and climate sensitivity are both much higher than generally assumed. They say that greater pollution controls and 'clean air' legislation will remove much of the aerosol cooling, and that if carbon dioxide levels are double their pre-industrial levels by 2100, a rise of 6 degrees can be expected. When this understanding is combined with predictions that parts of the natural carbon cycle after 2050 will reverse from being net absorbers to net emitters of carbon, they say that warming by 2100 may be as high as 10 degrees.

These findings have enormous implications. A long-term climate sensitivity of 6 degrees would mean that we have already passed the widely advocated 2-degree threshold of dangerous anthropogenic interference with the climate. It would, therefore, require us to find the means to engineer a rapid reduction of current atmospheric greenhouse gas even to restrict global warming to below 2 degrees—a target which we believe is, in any case, far too high.

A key question is whether the slow feedbacks have started to operate. In the case of the Greenland and Antarctic ice sheets, the data is already disturbing. One of the most important slow feedbacks to be considered is the reversing of the carbon cycle—as the oceans and soils take up less carbon dioxide—and the significant amounts of methane and carbon dioxide that are released by the permafrost.

Understanding how the carbon cycle works and how changes in the cycle will affect global warming are important in understanding the scale of action required to avoid catastrophic climate changes.

The carbon cycle is the flow and exchange of carbon in its various forms (including carbon dioxide, methane, and calcium carbonate) between the planet's four large, interconnected carbon reservoirs: the atmosphere (carbon dioxide); the oceans (carbon dioxide dissolved in seawater, carbon incorporated in living and non-living plants and animals, and methane trapped under pressure on the ocean floor); fossil organic carbon (coal, gas, and oil); and the land-surface biosphere (including soils, plants, and freshwater systems). Larger amounts are stored in the Earth's crust as rock carbonates, but these are relatively immobile.

The greatest carbon reservoir is the ocean, which contains about six times the amount of carbon that is stored in plants and soils. The fossil-fuels reservoir is of similar size to the land surface biosphere, while the atmospheric sink is the smallest.

Carbon flows between these reservoirs are driven by a variety of biological, physical, and chemical processes. Examples include extracting and burning fossil fuels; animal respiration; the exchange between the atmosphere and the oceans; drawing down carbon from the atmosphere by plant photosynthesis; and destroying forests by fire, land clearing, or decomposition.

Carbon reservoirs that absorb more carbon dioxide than they emit are called carbon *sinks* (as opposed to carbon *sources,* which emit more carbon dioxide than they absorb). The ocean is a carbon dioxide sink that responds rapidly to rising levels of atmospheric carbon, but not rapidly enough to meet the present need. The ocean water soaks up some of the additional carbon dioxide, and calcifying marine organisms absorb some of it (with subsequent burial in sea-floor sediments). Forests and grasslands also absorb some carbon dioxide by photosynthesis. Much of the carbon dioxide, however, remains in the atmosphere.

Many sinks governed by living organisms become less effective as the environment heats up. Though it has long been expected that the capacity of the Earth's carbon-drawdown mechanisms would decrease due to human activity and as a consequence of higher temperatures, changes already observed suggest that this is happening earlier than anticipated. The fraction of total human-caused carbon dioxide emissions that remain in the atmosphere has increased slowly with time — which implies a slight weakening of sinks, relative to emissions.

But some sinks may get to a point where they stop drawing down carbon and start emitting it instead. In 2000, a landmark study led by Peter Cox, then at the UK's Hadley Centre, found that about half the present emissions are being absorbed by the ocean and by land ecosystems. But this absorption is sensitive to the climate, as well as to atmospheric carbon dioxide concentrations. These two factors are creating a feedback loop, so that, under a 'business as usual' scenario, the terrestrial biosphere will only act as an overall carbon sink until about 2050, when it will fail and revert to being a carbon source. This slow feedback will increase temperatures by another 1.5 degrees by 2100.

Research published in October 2007 by Joseph Canadell, the executive director of the Global Carbon Project, confirmed that significant contributions to the growth of atmospheric carbon dioxide arise from the slow-down in the rate of absorption of natural sinks, or from 'a decrease in the planet's ability to absorb carbon emissions due to human activity'. According to Canadell: 'Fifty years ago, for every tonne of carbon dioxide emitted, 600kg were removed by land and ocean sinks. However, in 2006, only 550kg were removed per tonne and that amount is falling.' The data suggests that from 1959–2006 there was an implied decline of 10 per cent in the efficiency of natural sinks. Of the recent acceleration in the rise of atmospheric carbon dioxide levels, 18 per cent is attributed to the decreased efficiency of natural sinks.

Another key factor in this decreased efficiency has been identified by Peter Cox, now at Britain's Centre for Ecology and Hydrology in Dorset, who says that while plants are absorbing more carbon dioxide (because photosynthesis speeds up with warming), warming also encourages plant material in the soil to break down and release carbon dioxide. A lag between these events has seen the rise in carbon dioxide levels slowed for the last two decades; but science writer Fred Pearce says, 'Soon the biosphere will start to speed it up'. According to Cox, a possible surge of carbon dioxide into the atmosphere in 2003 is the first evidence of this process.

Cox spent years researching carbon cycles while at the Hadley Centre in Exeter, which has one of the world's most highly regarded climate-modelling systems. A summary of some of the centre's modelling work, published in 2005, included two startling graphs. In one, the amount of total carbon stored in the Amazon forest and soils shows a drop from around 70 billion tonnes of carbon in 2000 to just 20 billion tonnes of carbon by 2100. The second, using the same technique, compares vegetation and soil carbon levels in 2100 to those in 1850. While vegetation carbon had increased by about 60 billion tonnes of carbon by 2100, the amount of soil carbon had decreased by 130 billion tonnes.

The Amazon hosts a quarter of the world's species, and accounts for 15 per cent of land-based photosynthesis, as well as being an engine of regional and global atmospheric circulation and regional rainfall. Yadvinder Malhi of the Environmental Change Institute in Oxford led a team that concluded that the Amazon is warming at 0.25 degrees per decade, a rate twenty-five times faster than the temperature increase at the end of last ice age. There has already been an observed drying. Periods of recent drought in parts of the Amazon have increased the frequency of forest fires. With a total biomass store of 120 billion tonnes of carbon and predictions of large-scale drought in the eastern Amazon, the release of stored carbon by wildfires would be catastrophic.

Professor Guy Kirk of Britain's National Soil Resources Institute calculated that since 1978, the carbon lost by Britain's soil has increased by 13 million tonnes of carbon dioxide per year—more than the 12.7 million tonnes a year that Britain saved by cleaning up its industrial emissions as part of its commitment to the Kyoto Protocol. The loss is likely to be due to plant matter and soil organic material decomposing at a faster rate as temperatures rise—an effect that is expected to compound as temperatures increase. 'It's a feedback loop,' says Kirk. 'The warmer it gets, the faster it is happening.' It is thought that the terrestrial carbon sink will begin to convert to a carbon source at an increase of 2–3 degrees.

Bristol University researchers also argue that the previously unexplained surge of carbon dioxide levels in the atmosphere in recent years is due to more greenhouse gas escaping from trees, plants, and soils. Global warming is making vegetation less able to absorb the carbon pollution pumped out by human activity. Wolfgang Knorr believes that 'we could be seeing the carbon cycle feedback kicking in, which is good news for scientists because it shows our models are correct. But it's bad news for everybody else'. Another bad sign comes from

Canada's Manitoba region, where a study of a one-million-square-kilometre area of northern boreal forest found that the area is now releasing more greenhouse gases than it absorbs, because of an increased incidence of forest fires. This is consistent with predictions that climate change, by producing hotter and drier conditions, would lead to more fires. 'Those wildfires have caused this transition in the boreal forest from a carbon sink to a carbon source ... Climate change is what's causing the fires; if it was left unchecked, it could become a feedback,' says Tom Gower of the University of Wisconsin. A further consequence of wildfires is that more sunlight reaches the ground. This increases the rate of decomposition of organic matter, releases more carbon dioxide and, perhaps, contributes to the melting of the underlying permafrost.

Burning rainforests are also emitting hundreds of millions of tonnes of carbon dioxide each year. During the 2005–06 Amazon drought, thousands of square kilometres of land burned for months, releasing more than 100 million tonnes of carbon. Philip Fearnside of the National Institute for Research in the Amazon says that 'the threat of a "permanent El Niño" is to be taken very seriously ... Disintegration of the Amazon forest, with release of the carbon stocks in the biomass and soil, would be a significant factor in pushing us into a runaway greenhouse'. Daniel Nepstad, head of the Woods Hole Research Center's Amazon program, says:

[It is] not out of the question to think that half of the basin will be either cleared or severely impoverished just 20 years from now ... The nightmare scenario is one where we have a 2005-like year that extended for a couple of years, coupled with a high deforestation where we get huge areas of burning, which would produce smoke that would further reduce rainfall, worsening the cycle. A situation like this is very possible. While some climate modellers point to the end of the century for such a scenario, our own field evidence coupled with aggregated modelling suggests there could be such a dieback within two decades.

In October 2007, there were more than 10,000 points of fire across the Amazon, most of them having been set by ranchers to clear land. 'These fires are the suicide note of mankind,' says Hylton Murray-Philipson of the London-based charity Rainforest Concern.

A survey on tipping points, led by Tim Lenton of the University of East Anglia and published in early 2008, found that leading researchers estimated that there was a medium risk that the Amazon would be largely destroyed by 2050. (Regarding other potential tipping points, they also estimated a medium risk of the Indian summer monsoon destabilising within one year; the West African monsoon collapsing in 10 years; and the Arctic boreal forest dying in 50 years.)

Total carbon emissions from tropical deforestation are estimated at 1.5 billion tonnes of carbon a year, including illegal fires in Indonesia's vast peatlands, the haze from which regularly blankets Sumatra and Malaysia. Indonesia's peat swamps contain 21 per cent of the Earth's land-based carbon, and are now subject to increasing clearing, drying, and burning. During the 1997 El Niño event, an estimated 0.81–2.57 billion tonnes of carbon was released to the atmosphere as a result of burning peat and vegetation in Indonesia. This is equivalent to 13–40 per cent of the mean annual global-carbon emissions from fossil fuels. This burning also contributed greatly to the largest annual increase in atmospheric carbon dioxide concentration ever detected since records began in 1957.

New analysis of two decades of data from more than 30 sites also indicates that the ability of forests in the frozen north to soak up man-made carbon dioxide is weakening.

The melting of permafrost (permanently frozen soil, or soil below the freezing point of water) is another 'slow' feedback that is adding to global warming. As the Arctic warms, permafrost in the northern boreal forests, and further north in the Arctic tundra, is starting to melt. As it melts, its thick layers of thawing peat trigger the release of methane and carbon dioxide, both greenhouse gases.

With less than 1 degree of warming, Arctic ground that has been frozen for 3000 years is melting and producing thermokarst (a land surface that forms as ice-rich permafrost melts). Even under scenarios of modest climate warming, this could affect 10–30 per cent of Arctic lowland landscapes, and severely alter tundra ecosystems. As the permafrost thaws, lakes form and microbes convert the soil's organic matter into methane. The methane bubbles through the surface water into the atmosphere. In dry conditions, the warming soil also releases carbon dioxide.

A 2006 study found that Siberia's thawing wetlands are a significant, underestimated source of atmospheric methane, with lakes in the region growing in number and size, and emission rates appearing to be five times higher than previously estimated. The NCAR in Boulder predicts that half of the permafrost will thaw to a depth of 3 metres by 2050. As glaciologist Ted Scambos says: 'that's a serious runaway … a catastrophe lies buried under the permafrost.'

The western Siberian peat bog is amongst the fastest-warming places on the planet, and Sergei Kirpotin of Tomsk State University calls the melting of frozen bogs an 'ecological landslide that is probably irreversible'. One estimate puts methane releases from the current area of melting bog at 100,000 tonnes per day.

Russian Arctic climate researcher Sergei Zimov frames the gravity of the situation well: 'Permafrost areas hold 500 billion tonnes of carbon, which can fast turn into greenhouse gases ... The deposits of organic matter in these soils are so gigantic that they dwarf global oil reserves ... If you don't stop emissions of greenhouse gases into the atmosphere ... the Kyoto Protocol will seem like childish prattle.'

The ocean carbon-cycle feedback is also a significant slow-feedback contributor. Part of the decline in sink capacity comes from a decrease of up to 30 per cent in the efficiency of the Southern Ocean sink over the last 20 years. This decrease has been attributed to the strengthening of the winds around Antarctica, which enhances ventilation of natural, carbon-rich deep waters. Lead author Corinne Le Quéré of the University of East Anglia says:

> This is the first time that we've been able to say that climate change itself is responsible for the saturation of the Southern Ocean sink. This is serious. All climate models predict that this kind of 'feedback' will continue and intensify during this century. The Earth's carbon sinks—of which the Southern Ocean accounts for 15 per cent—absorb about half of all human carbon emissions. With the Southern Ocean reaching its saturation point, more carbon dioxide will stay in our atmosphere.

This finding follows pioneering work by CSIRO marine research scientists, including Stephen Rintoul and John Church, that seeks to understand how the Southern Ocean influences the climate system, its patterns of circulation, and the region's role in the global ocean-circulation system.

Measurements of the North Atlantic taken between the mid-1990s and 2005 found that, in the course of that decade, the amount of carbon dioxide in the water had reduced by half. It is suggested that warmer surface water was reducing the amount of carbon dioxide being carried down into the deep ocean. Lead researcher, Andrew Watson of the University of East Anglia, concludes: 'We suspect that it is climatically driven, that the sink is much more sensitive to changes in climate than we expected ... if you have a series of relatively warm winters, the ocean surface doesn't cool quite so much ... so the carbon dioxide is not being taken down into the deep water'. He warned that the process may fuel climate change: 'It will be a positive feedback, because if the oceans take up less carbon dioxide then carbon dioxide will go up faster in the atmosphere and that will increase the global warming.'

Satellite data gathered over the past ten years shows that the growth of marine algae, the basis of the entire ocean food chain, is being affected adversely by rising sea temperatures. Algae, the microscopic plants that permeate the oceans, remove up to 50 billion tonnes of carbon dioxide per year from the Earth's atmosphere. This system is as effective in removing carbon dioxide from the air as *all* plant life on the planet's land surface.

Jeff Polovina of Hawaii's National Marine Fisheries Service laboratory says that satellite imagery shows that green colouration (indicating chlorophyll life) in the middle of the ocean is fading away: 'The regions that are showing the lowest amount of plant life, which [are] sometimes referred to as the biological deserts of the ocean, are growing at roughly 1 to 4 per cent per year.' While such areas expanding are consistent with global warming scenarios, the rates of expansion already observed greatly exceed recent model predictions.

Increasing ocean acidification will also weaken marine life. This occurs as some of the carbon dioxide absorbed by the ocean reacts with water molecules to produce carbonic acid, which lowers the ocean's pH. The oceans are already 30 per cent more acidic than they were at the beginning of the Industrial Revolution, more than two centuries ago. If emissions continue at 'business as usual' rates, carbon dioxide levels in the oceans will rise so high that, by 2050, the ocean will be so acidic that current US water-quality standards would have to categorise it as industrial waste. Stanford University chemical oceanographer Ken Caldeira states that, if unabated, this could potentially cause the extinction of many marine species: 'What we're doing in the next decade will affect our oceans for millions of years … carbon dioxide levels are going up extremely rapidly, and it's overwhelming our marine systems.'

Waters around the Great Barrier Reef are also acidifying at a higher-than-expected rate. Ecosystem collapse caused by acidification will likely reduce marine biomass and, therefore, the capacity of the oceans to absorb carbon dioxide. Professor Malcolm McCulloch of the Australian National University says that, contrary to previous predictions, this acidification is now taking place over decades, rather than centuries: '[T]he new data on the Great Barrier Reef suggests the effects are even greater than forecast.'

Accumulating evidence suggests that slow feedbacks from oceans, soils, and permafrost are already affecting the climate system.

CHAPTER 6

# Most Species,
# Most Ecosystems

Martin Parry, co-chairman of one of the three IPCC working groups, told his audience at the launch of the full 2007 IPCC report on the impacts of global warming: 'We are all used to talking about these impacts coming in the lifetimes of our children and grandchildren. Now we know that it's us.' He said that destructive changes in temperature, rainfall, and agriculture were now forecast to occur several decades earlier than expected—and that means a huge threat to biodiversity.

As global temperatures rise, many species have to migrate towards the poles to stay in their habitable zones. If they can't migrate at sufficient speed, many species will be lost, and many ecosystems will degrade. During rapid change, such as the deglaciation and warming that occurred after the last ice age about 15,000 years ago, some widespread and dominant species became extinct when temperatures rose 5 degrees over a span of 5000 years. That is a rate of increase of 0.01 degrees per decade—20 times slower than today's rate of change.

Cagan Sekercioglu from Stanford University says that the IPCC's worst-case scenario to 2100, combined with extensive habitat loss, would result in the extinction of around 30 per cent of land bird species. With warming, birds will try to move to higher altitudes. Once the top of a mountain is reached, there is nowhere left to go. In the lowland tropics, where most bird species live, there can be no significantly higher slopes to which they can retreat.

The rate of change in temperature is also very important in determining the impact it will have, because many ecosystems and species are sensitive to small temperature changes. A study by Rik Leemans and Bas Eickhout found that if a 2-degree impact builds up slowly over 1000 years, most affected ecosystems are likely to adapt (most often by moving); but if the same rise happens in 50 years (0.4 degrees per decade), many ecosystems will deteriorate rapidly.

At 0.4 degrees per decade, the isotherms (bands of equal temperatures) will be moving towards the poles at about 120 kilometres per decade; at this rate of temperature change, most ecosystems will be torn apart. Interestingly, Australia's birds are moving south at a rate of 100–150 kilometres a decade with only half this rate of warming. Very fast-moving species will migrate with the temperature changes if they can survive in the ecosystems into which they move. Slow-moving species will not be able to keep up with the movement of their preferred temperature band and, unless they are tolerant of high temperatures and not dependent on species that have moved on, they will die out. At 0.4 degrees of change per decade, the isotherms are moving so fast that virtually all ecosystems will not be able to survive, and very large percentages of the dependent species will die out; yet this is the rate anticipated in some of the IPCC scenarios by mid-century, and few scenarios anticipate rates of less than 0.3 degrees per decade.

A 2007 study of the IPCC report's low- and high-emission scenarios, led by Dian Seidel of the NOAA in Washington, found that up to 39 and 48 per cent, respectively, of the Earth's terrestrial surface may experience novel and disappearing climates by 2100. Work published two years earlier projected the effects on 1350 European plant species under seven climate-change scenarios, and found that more than half could be vulnerable or threatened by 2080. The risk of extinction for European plants may be large, even in moderate scenarios of climate change.

Over the past 25 years, the area defined as 'climatologically tropical' has expanded to the north and south, away from the equator by about 2.5 degrees of latitude in each direction. This is equivalent to a rate of 110 kilometres per decade, and is greater than the IPCC's worst-case scenario of a total predicted shift of 2 degrees of latitude by 2100. This will disrupt the tropical–temperate geographic transition of ecosystems and, if maintained over a century timescale, it suggests that few of the affected ecosystems would adapt at the implied warming of greater than 0.3 degrees per decade.

Seidel and her team also found that the expanding equatorial belt has 'potentially important implications for subtropical societies and may lead to profound changes to the global climate system'. They argue that the pole-ward movement of large-scale atmospheric circulation systems such as jet streams and storm tracks 'could result in shifts in precipitation patterns affecting natural ecosystems, agriculture and water resources'. Of particular concern to them are subtropical dry belts that could affect water supplies, agriculture, and ecosystems over vast areas of the Mediterranean, the south-western United States, northern Mexico, southern Australia, southern Africa, and parts of South America.

For wooded tundra, an average of 27 per cent of the ecosystem would remain in place for a warming of 3 degrees in 100 years—or 0.3 degrees per decade, over a century timescale. At that rate, IPCC lead authors Rik Leemans and Bas Eickhout found that 'only 30 per cent of all impacted ecosystems ... and only 17 per cent of all impacted forests' can adapt. If the rate were to exceed 0.4 degrees per decade, all ecosystems would be quickly degraded, opportunistic species would dominate, and the breakdown of biological material would lead to even greater emissions of carbon dioxide. This would, in turn, increase the rate of warming.

With emissions already tracking higher than the worst scenario of the IPCC, we must conclude that 'business as usual' would see the destruction or degradation of most species and most ecosystems by mid-century.

# CHAPTER 7

# The Price of Reticence

Roger Jones is a CSIRO principal research scientist. On 10 December 2007, in Melbourne's *Herald Sun*, he issued this call for scientists to overcome their aversion to risk taking:

> Often, scientists do not like to release their results until they are confident of the outcome. Important decisions need to be made now and cannot wait another five to seven years. Scientists will have to leave their comfort zone and communicate their findings on emerging risks, even when scientific confidence in those findings may be low... Sometimes, it is worth taking some risks in the short term to avoid worse risks down the track. We have spent too long being risk-averse about short-term costs and ignored the benefits of avoiding long-term damages.

If only the IPCC would adopt such an attitude. Those turning to the 2007 IPCC reports for an up-to-date, authoritative view on global warming will find little of the real discussion of the events in the Arctic with which we started our story. The 2007 report is the IPCC's strongest call yet for governments, businesses, and communities to act immediately to reduce greenhouse emissions. But it is not enough, because it is based on outdated and incomplete data sets. The IPCC's four-year schedule for producing reports requires a submission deadline for scientific papers that is often two years, or more, before the report's final publication. What happens if there is significant new evidence, or dramatic events that change our understanding of the climate system, in the gap between the science reporting deadline and publication? They don't get a mention, which means that the IPCC report — widely viewed as the climate-change Bible — is behind the times even before it is released, though some new data is presented at forums.

On 28 January, just days before the release of the first of the IPCC's 2007 reports, the science editor of *The Observer*, Robin McKie, told of a serious disagreement between scientists over the report's contention that Antarctica will be largely unaffected by rising world temperatures:

> [M]any researchers believe it does not go far enough. In particular, they say it fails to stress that climate change is already having a severe impact on the continent and will continue to do so for the rest of century. At least a quarter of the sea-ice around Antarctica will disappear in that time, say the critics, though this forecast is not mentioned in the study. One expert denounced the [IPCC] report as 'misleading'. Another accused the panel of 'failing to give the right impression' about the impact that rising levels of carbon dioxide will have on Antarctica.

As McKie notes, the IPCC is, necessarily, a careful body.

Its reports involve the synthesis of many hundreds of pieces of research, and cooperation between many authors and contributors, such that only points that are considered indisputable by all of them are included: 'This consensus deflects potential accusations that the body might be exaggerating the threat to the planet. But the critics say it also means its documents tend to err too much on the side of caution.'

Under intense pressure from global-warming deniers, the IPCC has adopted some methods that have gone beyond being 'careful' and are now simply conservative.

Fred Pearce, writing in *New Scientist* on 10 February 2007, tells of an IPCC review process that was 'so rigorous that research deemed controversial, not fully quantified or not yet incorporated into climate models was excluded'. Pearce wrote: 'The benefit—that there is now little room left for sceptics—comes at what many see as a dangerous cost: many legitimate findings have been frozen out.' After interviewing many of the scientists involved, he described the process as 'a complex mixture of scientific rigour and political expediency [that] resulted in many of the scientists' more scary scenarios for climate change—those they constantly discuss among themselves—being left on the cutting room floor'.

The peer-review process for experimental science is conservative, insisting on verifiable, reproducible results. Peer-review can significantly delay the full publication of new findings. When research produces a range of outcomes with differing probabilities or risks, there is a tendency for the general reader, and even policy-makers, to be drawn to the middle position—or even to the low end of the range, which requires less action.

Wider uncertainties in climate science and the vulnerabilities of species to fast rates of temperature change are good examples, because they drive us to consider the worst outcomes—not just the scenarios that have average effects. Some of the high-impact scenarios considered by the IPCC to be 'extreme' are now looking quite likely.

Barrie Pittock says that uncertainties in climate-change science are inevitably large, due to inadequate scientific understanding, and to uncertainties in human agency or behaviour. He says:

> [Policies] must be based on risk management, that is, on consideration of the probability times the magnitude of any deleterious outcomes for different scenarios of human behaviour. A responsible risk-management approach demands that scientists describe and warn about seemingly extreme or alarming possibilities, for any given scenario of human behaviour (such as greenhouse gas emissions), even if they appear to have a small probability of occurring.

This, he says, is recognised in military planning, and is commonplace in insurance; the lesson for climate policy is that the object of policy advice must be to avoid unacceptable outcomes, not to determine the most apparent, likely, or familiar outcome.

Michael Oppenheimer and three fellow scientists agree, arguing that the emphasis on consensus has put the spotlight on expected outcomes, which then become anchored via numerical estimates in the minds of policy-makers; however, with the general credibility of the science of climate change established, they say it is now equally important that policy-makers understand the more extreme outcomes that consensus may exclude or downplay.

In the case of the Arctic, for example, it is clear that this has not been done. James Hansen laments:

> For the last decade or longer, as it appeared that climate change may be underway in the Arctic, the question was repeatedly asked: 'Is the change in the Arctic a result of human-made climate forcings?' The scientific response was, if we might paraphrase, 'We are not sure, we are not sure, we are not sure … Yup, there is climate change due to humans, and it is too late to prevent loss of all.' If this is the best that we can do as a scientific community, perhaps we should be farming or doing something else.

Pittock has described the limitations of the IPCC process:

> Vested interests harboured by countries heavily reliant on fossil fuels for industry and development, or for export, lead to pressure to remove worst-case estimates; scientists … tend to focus on 'best estimates', which they consider most likely, rather than worst cases that may be serious but which have only a small probability of occurrence; many scientists prefer to focus on numerical results from models, and are uncomfortable with estimates based on known but presently unquantified mechanisms; and due to the long (four-year) process of several rounds of drafting and peer and government review, an early cut-off date is set for cited publications (often a year before the reports appear).

Inez Fung at the Berkeley Institute of the Environment

says that for her research to be considered in the 2007 IPCC report, she had to complete it by 2004. 'There is an awful lag in the IPCC process,' she says, also noting that the special report on emission scenarios was published in 2000, and the data it contains were probably collected in 1998. 'The projections in the 2007 IPCC report [using the 2000 emission scenarios] are conservative, and that's scary', she says.

There is a widespread view that the more extreme an outcome, in a range of possibilities, the less likely it is to occur. This can underestimate the role of feedbacks in a non-linear world, and the evidence suggests that, in many cases, it is precisely the more extreme events that are coming true.

The data surveyed strongly suggests that, in many key areas, the IPCC process has been so deficient as to be an unreliable and, indeed, a misleading basis for policy-making. We need to look to processes that are not dogged by politics, and to a more up-to-date and relevant scientific knowledge base that integrates recent data and findings, expert comment, and the need to account for the most unacceptable, but scientifically conceivable, outcomes. On that basis we can build strategies that will at least give us a real chance to avoid the great dangers manifest in the climate system, of which humanity has become both master and victim.

The primary assumptions on which climate policy is based need to be interrogated. Take just one example: the most fundamental and widely supported tenet — that 3 degrees represents a reasonable maximum target if we are to avoid dangerous climate change — can no longer be defended. At less than a 1-degree rise, the Arctic sea-ice is headed for rapid disintegration; in all likelihood, triggering the irreversible loss of the Greenland ice sheet, catastrophic sea-level increases, and global warming from the albedo flip. Many species and ecosystems face extinction from the speed of shifting isotherms. Our carbon sinks are losing capacity, the seas are acidifying, and the tropical rainforests are fragile and vulnerable.

We have been lulled into a false sense of security by the stability of the climate during the Holocene period (the geological period that started 11,500 years ago, after the last glacial retreat, and which includes the whole period of human civilisation). Yet the period of ice ages and rapid deglaciations that occurred when the climate whipsawed between two states for millions of years is the usual mode. 'Abrupt change seems to be the norm, not the exception', says Will Steffen, head of the ANU's Fenner School of Environmental Science in Canberra. This is something we do not see, or do not want to see — and that incapacity means that, inevitably, abrupt changes, which our actions are now ensuring will occur, will be all the more devastating for our lack of foresight.

If we could start all over again, surely we'd say that we need to stabilise the climate at an equilibrium temperature that would ensure the continuity of the Arctic ice. This safe level has long since been passed. We should have acted rapidly to restore and maintain the Arctic ice cap, with a safe margin for uncertainty and error, as soon as we knew there was a problem. But, given what has happened, what choices do we have now?

PART TWO

# Targets

'We, the human species, are confronting a planetary emergency—a threat to the survival of our civilisation that is gathering ominous and destructive potential ... the Earth has a fever. And the fever is rising. The experts have told us it is not a passing affliction that will heal by itself. We asked for a second opinion. And a third. And a fourth. And the consistent conclusion, restated with increasing alarm, is that something basic is wrong. We are what is wrong, and we must make it right.'

— Al Gore, Nobel Peace Prize acceptance speech,
11 December 2007

# CHAPTER 8

# What We Are Doing

Something is wrong and we must make it right. This section explores the direction in which we must head to do so. How far must human greenhouse-gas emissions be reduced? What is a safe temperature zone? How do we get there?

In answering these questions, our first task is to understand what the greenhouse gases that we are pouring into the air have done, and what they are likely to do in the future.

In November 2006, *The New Yorker* reported on calculations by Ken Caldiera, from Stanford University, that 'a molecule of carbon dioxide generated by burning fossil fuels will, in the course of its lifetime in the atmosphere, trap a hundred thousand times more heat than was released in producing it'.

The quantity of carbon dioxide in the atmosphere and its persistence mean that it contributes more to global warming than any other product of human activity. Together with water vapour, methane, nitrous oxide, ozone, and trace gases, it maintains the Earth's greenhouse effect by trapping heat that radiates from the surface and, in doing so, keeps the surface temperature 33 degrees warmer than it would otherwise be.

Humans pour carbon dioxide into the air principally by processing and burning fossils fuels (coal, gas, oil, and its derivatives), and through the burning and decay of large amounts of organic material (as a result of changing land-use patterns and de-afforestation).

Human activity has increased the level of carbon dioxide in the air by 38 per cent from the 1750 pre-industrial level of 280 parts per million: by 2008, it was at 387 parts per million. According to UNESCO, this is the highest carbon dioxide concentration recorded in the past 600,000 years and, probably, the highest in the past 20 million years. What's more, the rate of increase has been at least ten, and possibly a hundred, times faster than at any other time in the past 400,000 years. So our species is creating energy imbalances in the climate system that are pushing the rate of change far more rapidly than at any time since modern humans began to walk the planet.

When carbon dioxide is added to the atmosphere, oceans absorb some of it, vegetation (through photosynthesis) absorbs some, and some is trapped in sediments or by chemical reactions with eroding rock. The portion that remains in the atmosphere, however, is so stable and long-lived that it continues to produce its greenhouse effect for hundreds, even thousands, of years. It is generally understood that if we stopped adding carbon dioxide to the air, the carbon cycle would gradually draw down the amount of atmospheric carbon dioxide and, slowly, over time, the temperature would decrease; but this may not happen over the short time relevant to our current predicament.

New research presents a very sobering picture. Ken Caldeira and his Stanford University colleague Damon Matthews have used climate modelling to demonstrate that the portion of carbon dioxide that remains in the air produces a temperature increase that persists for many centuries. In the terms of their study, this means for at least 500 years — which was as far into the future as their model was run.

They showed that current human-related carbon emissions will produce a temperature rise of 0.8 degrees that will persist for more than 500 years. In plain language, the carbon dioxide that we emit will keep the planet heated for many centuries, and the more we emit the higher the temperature over that period will be. The unavoidable bottom line, according to Matthews and Caldeira, is that if we want to stabilise temperatures, we must eliminate all carbon dioxide emissions. They show that 'stabilizing global temperatures at present-day levels [which are 0.8 degrees above the pre-industrial level] required emissions to be reduced to near-zero *within a decade* [our emphasis].' This is an important result to which we will return later in the story.

Methane quantities in the atmosphere are also increasing. Since 1750, they have increased by 150 per cent, and about half a billion tonnes of methane are added each year, mostly as a consequence of human activity. When the full impact of methane is accounted for, its heating effect—including the results of its interaction with other gases to form ozone in the lower atmosphere—is at least half that of human carbon dioxide emissions. (At low levels in the atmosphere, ozone contributes to smog and is polluting. This is distinct from its role at levels in the upper atmosphere, where it creates the protective 'ozone layer'.)

Drew Shindell of NASA's Goddard Institute for Space Studies estimates that methane may account for one-third of all climate warming from well-mixed greenhouse gases since the 1750s (carbon dioxide, methane, nitrous oxide, and halocarbons are known as well-mixed gases because their lifetime in the atmosphere is a decade or more): 'Control of methane emissions turns out to be a more powerful lever to control global warming than would be anticipated,' Shindell concludes.

Methane is produced when organic material decomposes in an anaerobic (oxygen-free) environment. Its main natural source is release from wetlands as a result of the decomposition of organic matter. Human activities that produce methane emissions include herding ruminant animals (cattle, sheep, and goats), growing rice, causing leakage during fossil fuel extraction, burning fossil fuels, causing gas to escape from waste landfills, and burning plant material.

Methane in the atmosphere chemically decomposes and loses its potency as a greenhouse gas in eight to 12 years, so it has a less persistent effect than carbon dioxide. If we significantly reduce our methane emissions, within a decade its effect as a heating agent and producer of lower-atmosphere ozone would be diminished, and a successful longer-term strategy to stop most human-caused methane emissions would take it off the agenda as a greenhouse gas of lasting concern.

Levels of nitrous oxide (known popularly as 'laughing gas') have also increased—they are up by 16 per cent since 1750. While relatively small in concentration, the gas has an effect three hundred times more powerful than carbon dioxide, making its overall contribution to global warming about one-tenth that of carbon dioxide.

The majority of nitrous oxide is emitted naturally from tropical soils and oceans. The human activity that produces most nitrous oxide is agriculture (through the use of fertilisers), but jet engines, some industrial processes, and cars with catalytic converters that burn fossil fuels also contribute to its production. The gas persists in the atmosphere for about 120 years before being broken down by the effect of sunlight; nonetheless, it is slowly accumulating in the air as a consequence of additional human-caused emissions.

A number of other gases (known as 'trace gases') that are emitted in smaller quantities from industrial processes — including hydroflourocarbons, sulfur hexafluoride, and perfluorocarbon — contribute to global warming, but on a smaller scale. These gases, together with carbon dioxide, methane, and nitrous oxide, are known as the 'Kyoto gases', because they are defined under the agreement to control emissions that was established by the Kyoto Protocol in 1997.

While human activity since 1750 has raised the carbon dioxide level by 38 per cent to 387 parts per million in 2008, the effect of all the Kyoto gases together is calculated to be equivalent to 455 parts per million of carbon dioxide.

Human activities also contribute to the greenhouse effect by releasing non-gaseous substances such as aerosols, which are small particles that exist in the atmosphere. Aerosols include black-carbon soot, organic carbon, sulphates, nitrates, as well as dust from smoke, manufacturing, windstorms, and other sources.

Aerosols have a net cooling effect because they reduce the amount of sunlight that reaches the ground, and they increase cloud cover. This effect is popularly referred to as 'global dimming', because the overall aerosol impact is to reduce, or dim, the sun's radiation, thus masking some of the effect of the greenhouse gases. This is of little comfort, however, because aerosols, or airborne particle pollution, last only about ten days before being washed out of the atmosphere by rain; so we have to keep putting more into the air to maintain the temporary cooling effect. Unfortunately, the principal source of aerosols is the burning of fossil fuels, which causes a rise in carbon dioxide levels and global warming that lasts for many centuries. The dilemma is that if you cut the aerosols, the globe will experience a pulse of warming as their dimming effect is lost; but if you keep pouring aerosols together with carbon dioxide into the air, you cook the planet even more in the long run.

There has been a necessary effort to reduce emissions from some aerosols because they cause acid rain and other forms of pollution. However, in the short term, this is warming the air as well as making it cleaner.

The total effect of aerosol cooling is generally estimated to be less than 1 degree; however, work by Nicolas Bellouin and a team from the UK Met Office that was published in *Nature*, in December 2005, found that the cooling effect of aerosols at around 1.4 degrees is much greater than most current climate models estimate. The corollary is that, since aerosols emissions continue to decline, they will be less able to create a cooling effect, and therefore future global warming from greenhouse gases will be greater than presently indicated.

This view is consistent with the idea that climate sensitivity is higher than is generally taken to be the case, as we discussed in Chapter 6. This has led Meinrat Andreae of the Max Planck Institute for Chemistry in Mainz, Germany, to conclude that a doubling of pre-industrial carbon dioxide levels by 2100 would produce a 6-degree increase, which would 'be comparable to the temperature change from the previous ice age to the present [and] so far outside the range covered by our experience and scientific understanding that we cannot with any confidence predict the consequences for the Earth'. Andreae's collaborator, Chris Jones, warns: 'Now we are taking our foot off the brake, but we don't know how fast we will go. Because we don't know exactly how strong the aerosol cooling has been, we do not know how strong the greenhouse warming will be.'

While most aerosols act to cool the planet, one component, black carbon, has the opposite effect. Black carbon particles (which are created by burning vegetation; heating with coal; diesel combustion5; and cooking with solid fuels, such as wood and cow dung) act in a similar manner to the greenhouse gases by trapping heat radiating away from the Earth's surface, and by changing the reflective properties of ice-sheets. A study by Scripps Institution of Oceanography atmospheric scientist V. Ramanathan and University of Iowa chemical engineer Greg Carmichael has found that soot and other forms of black carbon may have a heating effect greater than any other greenhouse gas, and 60 per cent stronger than that of carbon dioxide. This is three times the effect estimated by the IPCC.

This is good news, in a roundabout way. This discovery means that strong action to cut black-carbon emissions could balance some of the cooling losses that occur when other aerosols produced by burning fossils fuels are reduced.

Together, current levels of greenhouse gases that are caused by human activity are working to produce the following global warming:

| | |
|---|---|
| Long-term effect of the present level of carbon dioxide | 1.4°C |
| *Plus* the effect of non-carbon-dioxide levels of Kyoto gases (methane, etc.) | 0.7°C |
| *equals* the total impact of all Kyoto gases | 2.1°C |
| *minus* thermal inertia (heat being used to warm the oceans) | 0.6°C |
| *minus* the short-term net cooling effect of aerosols | 0.7°C |
| *equals* today's warming | 0.8°C |

To add another level of complexity, all these estimates are based on a climate sensitivity of 3 degrees for fast feedbacks, which is the middle of the range used by the IPCC. As we saw in Chapter 6, this 3-degree estimate is reasonable in the short term, but there is strong evidence that the figure is double that, at 6 degrees, when all the long-term consequences and slow feedbacks are accounted for.

CHAPTER 9

# Where We Are Headed

To help think about possible future trajectories of human-produced greenhouse gases, the IPCC has developed six sets of scenarios, each of which makes different assumptions about future emissions, land use, technologies, and forms of economic development. The scenarios range from those that assume large reductions in greenhouse-gas emissions to those that assume a world of 'business as usual' practices and, as such, imagine the most pessimistic, fossil-fuel-intensive emissions future. The current IPCC scenarios were prepared for the panel's 2001 report and are now almost a decade old, lagging well behind reality. As Roger Jones of the CSIRO says, 'At the time of their release in 2000, [the scenarios] were state-of-the-art … Now, the world is growing faster and is richer than the scenario authors assumed.'

According to the most recent IPCC report, human-caused carbon dioxide emissions increased 70 per cent between 1970 and 2004, and are rising at an even faster rate now. Their annual increase jumped from an average of just over 1 per cent for the period from 1990–1999 to more than 3 per cent from 2000–2004. The actual growth rate of carbon-dioxide emissions since 2000 is greater than growth rates for the most fossil-fuel-intensive of the IPCC emissions scenarios.

A study led by the CSIRO's Michael Raupach, co-chair of the Global Carbon Project, has found that no region is effectively decarbonising its energy supply. Raupach says that a major driver accelerating the growth rate in global emissions is that we're now burning more carbon for every dollar of wealth we create: 'In the last few years, the global use of fossil fuels has actually become less efficient. This adds to pressures from increasing population and wealth.'

In Australia, Raupach says, carbon emissions have grown at about twice the global average during the past 25 years, and have almost doubled the growth rate of emissions in the United States and Japan. He believes that because 'emissions are increasing faster than we thought … the impacts of climate change will also happen even sooner than expected'.

According to the October 2007 World Bank report *Growth and Carbon Dioxide Emissions: how do different countries fare?*, Australia increased its carbon dioxide emissions by 38 per cent between 1994 and 2004, to become the sixth-highest per capita emitter (on a base that excludes land use, land-use change, and forestry). Australia's emissions-increase was more than the total of Britain, France, and Germany which, combined, have a population ten times that of Australia.

The rising rate of global carbon dioxide emissions is reflected in a larger annual increase in the level of carbon dioxide in the air. The average increase of 1.5 parts per million for the period from 1970–2000 has jumped to 2.1 parts per million since 2001. NASA's James Hansen told the *Independent* in January 2007 that 'if we go another ten years, by 2015, at the current rate of growth of carbon dioxide emissions, which is about 2 per cent per year, the emissions in 2015 will be 35 per cent larger than they were in 2000'. He says that this would take the emissions scenarios necessary to avoid dangerous climate change beyond reach.

Atmospheric carbon dioxide levels are now rising faster than at any time in the past 800,000 years. The level rose 30 parts per million over the past 17 years; yet ice cores drilled in Antarctica show that in the past million years, prior to recent times, the fastest increase of carbon dioxide was 30 parts per million over a period of a thousand years.

The increasing use of energy is also going to increase emission levels. In 2004, the International Energy Agency (IEA) projected that annual carbon dioxide emissions by 2030 would be 63 per cent higher than in 2002. According to the European Union's 2007 *World Energy Technology Outlook*, 'business as usual' will see global energy use more than double by 2050, with 70 per cent of the increase coming from fossil fuels. The report assumes that energy efficiency will almost double, in order to support an economy four times larger than today. The result would be a carbon dioxide concentration in the atmosphere of 900–1000 parts per million by 2050. It says: 'This value far exceeds what is considered today as an acceptable range for stabilisation of the concentration.' The conclusion is that carbon emissions cuts will come too late to avert 'runaway' climate change if current policy trends continue, and that this would happen despite a 'massive' growth in renewable energy after 2030, including rapid deployment of new technologies, such as offshore wind.

While the IEA predicts annual growth in global power consumption of 3.3 per cent per year to 2015, a study by

Oxford Economics analysts shows that, when trends in developing countries are studied in more detail, the rate would be even higher, at 5 per cent.

Increasing energy use and rates of greenhouse-gas emissions mean only one thing: it will get hotter, quicker. The IPCC's conservative estimate is a rise of 4 degrees by 2100 for the most pessimistic 'business as usual' scenario, yet our emissions are currently rising faster than this scenario envisages. The ten warmest years on record have all occurred since 1995, and one study predicts a 0.3-degree increase for the period from 2004–2014 alone.

Before the Arctic big melt of 2007, Hansen and his colleagues, by comparing sea-surface temperatures in the Western Pacific with historical climate data, suggested that this critical ocean region, and probably the planet as a whole, is 'approximately as warm now as at the Holocene maximum [the period of the highest temperature within the last 11,500 years] and within *one degree of the maximum temperature of the past million years*' [our emphasis]. They conclude that global warming 'of more than one degree, relative to 2000, will constitute "dangerous" climate change as judged from likely effects on sea level and extermination of species'.

Rates of warming since the mid-19th century are higher than those of the last ice age by more than a factor of ten, increasing to a factor of twenty from the mid-1970s. The atmosphere is now heating up more quickly than modern humans have ever experienced. 'We really are in a situation where we don't have an analogue in our records,' says Eric Wolff from the British Antarctic Survey. According to Wolff, it is generally accepted that at some stage a 'step change' or 'tipping point' is reached, after which global warming accelerates exponentially. According to new evidence, he says, 'we could expect that tipping point to arrive in ten years' time.' Recent observations from the Arctic, and their implications for the Greenland ice sheet and sea-level rises, suggest that we may have already passed that point.

When accepting the WWF Duke of Edinburgh Conservation Medal in November 2006, James Hansen told his audience that the human race must begin to move its energy systems in a fundamentally different direction within about a decade, or 'we will have pushed the planet past a tipping point beyond which it will be impossible to avoid far-ranging undesirable consequences'. He warned that global warming of 2 to 3 degrees above the present temperature would produce a planet without Arctic sea-ice; a catastrophic sea-level rise of around 25 metres; and a super-drought in the American west, southern Europe, the Middle East, and parts of Africa. Such a scenario, he says, 'threatens even greater calamity, because it could unleash positive feedbacks such as melting of frozen methane in the Arctic, as occurred 55 million years ago, when more than 90 per cent of species on Earth went extinct'.

The ANU's Will Steffen argues that the Earth's climate system 'is highly non-linear and is prone to abrupt changes, threshold effects and irreversible changes' in a human time frame, so that very small changes in a forcing factor 'can trigger surprisingly large and sometimes catastrophic changes in a system ... [and] propel the Earth into a different climatic and environmental state'. Examples he cites include 'the rapid disintegration of the large ice sheets on Greenland and Antarctica or large-scale and uncontrollable feedbacks in the carbon cycle: activation of methane clathrates [frozen water and methane] buried under sediments on the ocean floor, the rapid loss of methane from warmer and drier tundra ecosystems, increasing wildfires in the boreal and tropical zones, the conversion of the Amazon rainforest to a savannah and the release of carbon dioxide from warming soils'. Once we cross critical thresholds and trigger these processes, Steffen says no policy or management approach could slow, or reverse, the process.

Hansen agrees. He says the tipping point occurs when the climate state is close to triggering very strong positive-feedback effects, so that a small perturbation can cause large climate change.

Today, the Arctic sea-ice, the West Antarctic ice sheet, and the Greenland ice sheet can provide such feedbacks. Little additional forcing is needed to trigger these feedbacks, because of the warming that is already in the pipeline. Hansen concludes wryly: 'We have to be smart enough to understand what is happening early on.'

Tony Blair and his Dutch counterpart Jan Peter Balkenende told European leaders in 2006 that, 'without further action, scientists now estimate we may be heading for temperature rises of at least 3 to 4 degrees above pre-industrial levels … We have a window of only ten to 15 years to avoid crossing catastrophic tipping points. These would have serious consequences for our economic growth prospects, the safety of our people, and the supply of resources—most notably, energy.'

This statement was made before the imminent loss of the Arctic sea-ice, and the consequences of that loss, were as clear as they are today. When that event is taken into account, the ten-to-15-year window looks to be closed already.

# CHAPTER 10

# Target 2 Degrees

Climate change is already dangerous. The signs are evident globally: in the polar north; in Darfur's famine; in Australia's depleted Murray–Darling River system; in the collapse of ecosystems across the globe; in the 2007 mega-fires in Greece and California; in the coral stress in the Caribbean and in Australia's Great Barrier Reef; in widespread species losses; in changing monsoon patterns; in the destruction of low-lying communities; and in regional food-production stress. Our world is already at the point of failing to cope. The UN's emergency relief coordinator, Sir John Holmes, warned that 12 of the 13 major relief operations in 2007 were climate related, and that this amounted to a climate-change 'mega disaster'.

Global warming is now close to 1 degree. Many of the results that were forecast are already coming true.

At a warming of just 1 degree over pre-industrial levels, it was predicted that the Amazon would be drying, and increasingly drought and fire affected. During the 2005 drought, some tributaries ran dry; in 1998, forest fires generated by El

Niño conditions poured almost half a billion tonnes of carbon into the air—more than 5 per cent of global greenhouse-gas emissions for that year. The Amazon, responsible for more than 10 per cent of the world's terrestrial photosynthesis, is currently near its critical-resiliency threshold.

In the US, it was expected that a 1-degree rise would result in California and the Great Plains states becoming subject to mega-droughts and desertification: a new and permanent 'dust bowl', similar to those seen between 1000 and 1300 AD during the Medieval Warm Period, when devastating, epic droughts hit the plains, and whole Native American populations collapsed. This predicted drying is also occurring.

At a warming of just 1 degree, the North Queensland Wet Tropics rainforest will be an environmental catastrophe waiting to happen, according to Steve Williams, a James Cook University senior research fellow. Just 1 degree is likely to reduce the area of this World-Heritage-listed Queensland highland rainforest by half. As predicted, the Barrier Reef is already subject to regular bleaching (loss of colour due to loss of algae), and is now facing extinction: a survey showed that 60–95 per cent of it was bleached in 2002. This is the case with most coral reefs around the world.

At 1 degree of warming, it was also expected that world cyclones would be more severe, and that small island states would be abandoned as seas rose. This is happening.

As predicted for a 1-degree rise, ice sheets around the world are suffering severe losses; and as permafrost melts, landslides in the European Alps are already becoming serious. The Mount Kilimanjaro ice cap, which has been intact for at least 11,000 years, is well on the way to disappearing, with an 80 per cent loss in the last hundred years, and the rest predicted to be gone between 2015 and 2020 as surrounding forests die off.

Britain's Hadley Centre calculated that warming of just 1 degree would eliminate fresh water from a third of the world's land surface by 2100, worsening a water crisis that seems already to be a permanent new part of life in many parts of the world.

All of these effects have occurred with a 1-degree warming; yet the most commonly used definition of dangerous climate change is linked to a 2-degree warming threshold, and its corollary, suggested by Sir Nicholas Stern, among others, that our target should be for a 60 per cent reduction in greenhouse-gas emissions by 2050.

It is important to understand how these numbers achieved such prominence in the climate debate. The first goal set by a forum of international significance was in 1988. The International Conference of the Changing Atmosphere in Toronto advocated a 20 per cent reduction of 1988 carbon dioxide levels by 2005.

In 1990, the first IPCC Scientific Assessment Report pointed out—for educational rather than policy purposes—that it would require a 60–80 per cent cut in emissions if carbon dioxide emissions were to be stabilised at the then-current level of around 350 parts per million. This guesstimate was superseded four years later when CSIRO scientist Ian Enting and his colleagues reported the results of ten world climate models, eight of which showed that the reductions required to stabilise the atmosphere at 350 parts per million of carbon dioxide would likely be more than 100 per cent—that is, carbon dioxide emissions would need to be completely eliminated, and carbon would need to be taken out of the air, for 50–90 years.

By 1997, however, governments were not thinking of cuts on anything like this scale. This was reflected in the Kyoto Protocol's target for developed-country emissions to be only 5 per cent less than the 1990 level by 2012. Achieving this target would result in annual additions of carbon dioxide to the atmosphere of around of 6 billion tonnes, which would not stabilise greenhouse gases in the air for hundreds of years, and would likely see the level of carbon dioxide climb past 1000 parts per million—more than three times the highest level known in the last million years.

In 2000, realising that the Kyoto cuts were inadequate, Britain's Royal Commission on Environmental Pollution recommended that if greenhouse gases were to be stabilised at 550 parts per million carbon dioxide equivalent, emissions from Kyoto Annex I [developed] nations would need to be reduced to 60 per cent below 1998 levels by 2050. It was argued that this target was needed if the world was to avoid a 2-degree warming. Six years later, however, the Stern Review published data indicating that if the atmosphere was stabilised at 550 parts per million carbon dioxide equivalent, there would be a 99 per cent chance of exceeding a 2-degree warming; so the 2-degree target was then shifted to be associated with an emissions cap of 450 parts per million. To stabilise the atmosphere at 450 parts per million of carbon dioxide equivalent, a reduction in emissions to at least 80 per cent less than the level in 1990 is required; yet the target of 60 per cent by 2050 remained a popular government policy long after it was shown to be inadequate.

If we accept that the present rise of 0.8 degrees (with more warming in the pipeline) is already dangerous, we can no longer assume that we have another 40 years in which to reduce emissions to 60–80 per cent below 1990 levels, as argued by those advocating a higher temperature cap of 2 degrees. Nevertheless, it is worth looking at the proposed emission scenarios—the scale and speed of emission reductions—necessary to achieve a 2-degree target, because they demonstrate that even this inadequate target will not be achieved by governments acting in their 'business as usual' mode.

The European Union, the IPCC, and the International Climate Change Taskforce, among many others, propose a temperature cap of 2 degrees to avoid 'dangerous anthropogenic interference with the climate system'. For a 2-degree cap, research finds that, in the long run, the Kyoto-defined greenhouse gases need to drop below 400 parts per million, and they need to be significantly less if the risk of overshooting the target is to be low.

Malte Meinshausen of the Potsdam Institute for Climate Impact Research in Germany calculates: 'Our current knowledge about the climate systems suggests that only stabilization around or below 400 parts per million carbon dioxide equivalence will likely [85 per cent probability] allow us to keep global mean temperature levels below 2 degrees in the long term.'

Similarly, Simon Rettalack of the Institute for Public Policy Research in the UK says that to have an 80 per cent chance of keeping global average warming below 2 degrees, 'greenhouse-gas concentrations would need to be prevented from exceeding 450–500 parts per million carbon dioxide equivalent in the next 50 years and thereafter should rapidly be reduced to about 400 parts per million carbon dioxide equivalent'.

Compared to 'business as usual' scenarios, 2-degree scenarios are characterised by a very sharp turnaround in emissions — falling to, or below, half of the 1990 level by 2050 — and then declining towards zero. The downward-sloping curves are so steep that they can only be called crash programs.

There are large uncertainties about the relationship between the level of greenhouse gases in the atmosphere and the long-term temperatures that will accompany them. This necessitates the expression of ranges, or probabilities, of outcomes. The Stern Review, using calculations by the Hadley Centre in the UK, shows that, in the long term, greenhouse-gas levels of 400 parts per million carbon dioxide equivalent have a 33 per cent probability of exceeding 2 degrees; a 3 per cent chance of passing 3 degrees; and a 1 per cent chance of exceeding 4 degrees.

Because today's carbon dioxide level alone is close to the long-term cap of 400 parts per million of carbon dioxide equivalent and emissions are still rising, the 2-degree strategies depend on 'peak and decline'. This means that the maximum target is breached, but because of the time lag between the increase in greenhouse-gas concentrations and the increase in temperature, there is an opportunity to lower emissions and have greenhouse gases drawn down by the carbon cycle before the theoretical maximum temperature is reached.

Meinshausen describes the process:

Fortunately, the fact that we are most likely to cross 400ppm [parts per million] CO2eq [carbon dioxide equivalent] level in the near-term, does not mean that our goal to stay below 2°C is unachievable. If global concentration levels peak this century and are brought back to lower levels again, like 400ppm, the climate system's inertia would help us to stay below 2°C. It's a bit like cranking up the control button of a kitchen's oven to 220°C (the greenhouse gas concentrations here being the control button). Provided that we are lowering the control button fast enough again, the actual temperature in the oven will never reach 220°C.

For a 70–90 per cent chance of staying below 2 degrees, Meinshausen maps an 'initial peak at 475 parts per million carbon dioxide equivalent', leading to the long-term return to '400 parts per million carbon dioxide equivalent'.

'Peak and decline' assumes that emissions will eventually be cut to below the Earth's net carbon-sink capacity; it assumes that there is a mechanism operating to remove the excess carbon dioxide from the air to lower the level of greenhouse gases from the peak, before their full force is felt. But if the weakening of the carbon sinks, as predicted and observed, is sufficiently large, this drawdown effect will not be strong enough. In this case, unless the natural carbon sinks are supplemented by a human-organised carbon dioxide drawdown of atmospheric carbon on a huge scale, 'peak and decline' will be a failed strategy, and atmospheric greenhouse gases will be stranded at a far higher level than planned.

A number of researchers have attempted to estimate the level that emissions would need to be cut to stabilise at a 2-degree rise with a carbon dioxide equivalent of 400 parts per million. Research by Paul Baer and Michael Mastrandrea found that global emissions of carbon dioxide would need to peak between 2010 and 2013; achieve a maximum annual rate of decline of 4–5 per cent some time between 2015 and 2020; and fall to about 70–80 per cent below 1990 levels by the middle of the century. At the same time, similarly stringent reductions in the other greenhouse gases would need to occur.

Meinshausen says: 'To avoid a likely global warming of more than 2°C and all its consequences, global emissions would need to be reduced significantly, i.e. around 50 per cent by 2050. Per-capita greenhouse-gas emissions would need to be reduced by around 70 per cent, so that global emissions could be halved despite the globally increasing population.'

Using mid-range climate sensitivity, a team led by the Center for International Climate and Environmental Research's Nathan Rive found that even getting to a 50 per cent chance of preventing more than 2 degrees of warming would require a global cut of 80 per cent by 2050, if total emissions were to peak in 2025. For a lower risk of failure than 50 per cent, the emission cuts would need to be substantially higher.

Now comes the crunch for Australia. Because Australian emissions are five times the global average, and the world population will be half as large again by 2050, these scenarios require Australian per-capita emissions to be cut by at least 95 per cent by 2050—a proposition currently rejected by Australia's Rudd government.

But how worthwhile will those cuts be, in any case, if 2 degrees is too much? A rise of 2 degrees over pre-industrial temperatures will initiate climate feedbacks in the oceans, on ice-sheets, and on the tundra, taking the Earth well past significant tipping points. As we have seen, likely impacts include large-scale disintegration of the Greenland and West Antarctic ice-sheet; the extinction of an estimated 15–40 per cent of plant and animal species; dangerous ocean acidification; significant tundra loss; increasing methane release; initiation of substantial soil and ocean carbon-cycle feedbacks; and widespread drought and desertification in Africa, Australia, Mediterranean Europe, and the western USA. At 2 degrees, Europe is likely to be hit by heatwaves every second year, much like the one in 2003 that killed up to 35,000 people, caused US$12 billion of crop losses, reduced glacier mass, and resulted in a 30 per cent drop in plant growth that added half a billion tonnes of carbon to the atmosphere.

At 2 degrees of warming, the summer monsoons in northern China will fail, and agricultural production will fall in India's north as forests die back and national production falls. Flooding in Bangladesh will worsen as monsoons strengthen and sea levels rise. In the Andes, glacial loss will reach 40–60 per cent by 2050, reducing summer run-off and causing horrendous water shortages in South American nations. At 2 degrees, California will see a decline in the snowpack of one-third to three-quarters, with a loss of up to 70 per cent in the Northern Rockies, which will devastate regional agriculture as melt run-off declines. Changing climate will have a severe impact on world food supplies: in central and South America, maize losses are projected for all nations but two. In 29 African countries, crop failure and hunger are likely to increase.

After a careful reassessment of climate sensitivity and the climate history data, James Hansen and seven co-authors are now suggesting that the tipping point for the presence, or absence, of *any* substantial ice-sheets on Earth seems to be at around 425 parts per million (plus or minus 75 parts per million) of carbon dioxide. This means that the carbon dioxide levels often associated with a 2-degree temperature rise may also be the tipping point for the total loss of all ice sheets on the planet, with an eventual sea-level rise of 70 metres.

Despite the catastrophic consequences of a 2-degree warming, the European Union and the International Climate Change Taskforce, among many others, have set 2 degrees as the target towards which the world should aspire.

But if 2 degrees spells disaster, what will the new 'business as usual' target of 3 degrees bring?

# CHAPTER 11

# Getting the Third Degree

The rapid Arctic melt consigns the widely advocated 2-degree-warming cap—always an unacceptable political compromise—to the policy dustbin. Scientific evidence shows it is too high and would be a death sentence for billions of people and millions of species.

In late 2007, Australian government advisor and former chief of CSIRO Atmospheric Research Graeme Pearman wrote:

> The global climate-science community has indicated that changes of planetary temperature of even one-to-two degrees have the potential to bring about significant global exposures to coastal erosion, sea-level rise, water supply and extreme climatic events, to name but a few. The potential number of humans impacted by a 2-degree change may count in the hundreds of millions. The European Union has already set a target of maximum warming of 2 degrees in the belief that warming beyond this represents an unreasonable risk of 'dangerous' climate

change. Such a change in the average global temperature might be regarded by many as small, but it has the capacity to culminate in major consequences, something that scientists feel is still under-appreciated in both public and private policy development.

Despite the dangerous consequences of 2 degrees of warming, we are now being asked by politicians to consider a 3-degree warming cap, because they consider the 2-degree target to be too great a challenge for their 'business as usual' mode of operation.

Thanks to recent developments in paleoclimatology, we have some insights into what a 3-degree world might be like. In the Pliocene, three million years ago, temperatures were 3 degrees higher than our pre-industrial levels. In that era, the northern hemisphere was free of glaciers and ice sheets, beech trees grew in the Transantarctic Mountains, sea levels were 25 metres higher than they are today, and atmospheric carbon dioxide levels were 360–400 parts per million, very similar to today. There are also strong indications that, during the Pliocene, permanent El Niño conditions prevailed. Rapid warming today is already heating up the western Pacific Ocean, a basis for a coming period of 'super El Niño'.

At 2–3 degrees of warming above pre-industrial levels, and perhaps at much lower warming levels than that, the Amazon rainforest will also suffer devastating damage: its plants, which produce 10 per cent of the world's terrestrial photosynthesis, have no evolved resistance to fire, and the warming may result in it becoming savannah. Further, the carbon released by the forests' destruction will be joined by carbon release from the world's warming soils. This will boost global temperatures by

1.5 degrees, on top of the warming of around 4 degrees by 2100 that is projected to occur if we keep to the current fossil-fuel intensive path.

The climate-change model at the UK's Hadley Centre predicts that the chances of an Amazon forest drought would rise from 5 per cent now, to 50 per cent by 2030, and to 90 per cent by 2100. Four or five consecutive years of drought would probably dry out areas of the Amazon sufficiently for wildfires to destroy much of it.

The collapse of the Amazon is part of the full reversal of the carbon cycle that is projected to happen at around 3 degrees of warming—a view confirmed by a range of researchers using carbon-coupled climate models. A 3-degree warming would see significantly large areas of the Earth's terrestrial environment rendered uninhabitable by drought and heat. Rainfall in Mexico and Central America is projected to fall by 50 per cent. Southern Africa would be exposed to perennial drought, and a huge expanse of land centred on Botswana could see a remobilisation of sand dunes, as is predicted to happen in the western US even sooner. The Rockies would be snowless, and water flows in the Colorado River would fail one year in two. Drought intensity in Australia could triple, and World Heritage ecosystems would severely degrade or die, while hurricanes could increase in power by half a category above today's top-level Category Five.

With such extreme weather conditions, world food supplies will be devastated. This could mean billions of refugees moving towards the mid-latitudes from areas of famine and drought in the sub-tropics. As rising temperatures cause the Himalayan ice sheet to melt, long-term water-flows into Asia's great rivers and breadbasket valleys—including the

Indus, Ganges, Brahmaputra, Mekong, Yangtse, and Yellow rivers — will fall dramatically. It has been predicted that if global temperatures rise by 3 degrees, which is becoming the unofficial target for some governments of richer nations, water flow in the Indus would drop by 90 per cent by 2100. But the loss of the Himalayan ice-sheet now looks likely to occur at well less than a 3-degree rise. Recent estimates are that the Himalayas may be completely ice-free before 2050, or even sooner. The lives of two billion people are at stake.

For all this, 3 degrees is the cap effectively being advocated by Australia's Labor government. In its 2007 pre-election policy, Labor advocated a 60 per cent reduction by 2050 in Australian emissions from 2000 levels. Environment Minister Penny Wong then reaffirmed this, in February 2008, when she tersely rejected suggestions from the Garnaut Review that the cut might need to be 90 per cent.

The goal of a 60 per cent reduction in emissions by 2050 (known as '60/2050') for fully developed nations was first formally articulated by a major organisation in 2000, when it was recommended by the UK Royal Commission on Environmental Pollution. The Rudd government's policy statement makes reference to this; however, the core idea — to make a 60 per cent cut in carbon dioxide emissions compared to 1990 levels — had been given prominence a decade earlier, in the first science assessment of the IPCC. This was not presented as a goal, as such — it was provided by the scientists to help policy-makers take in the scale of the challenge.

The immediate source of inspiration for Labor's 60/2050 target appears to be the Stern Review, which advocated a 3-degree target. In the report, Stern stated that constraining greenhouse-gas levels to 450 parts per million carbon dioxide equivalent 'means around a fifty-fifty chance of keeping global increases below 2 degrees above pre-industrial [and it] is unlikely that increases will exceed 3 degrees'; but, he said, keeping to this is 'already nearly out of reach', because it means 'peaking in the next five years or so and dropping fast'. It would require immediate and strong action, which Stern judged to be neither politically likely nor economically desirable.

Instead, Stern pragmatically said the data 'strongly suggests that we should aim somewhere between 450 and 550 parts per million carbon dioxide equivalent'. However, his policy proposals demonstrate that he has the higher figure in mind as a practical goal: 'It is clear that stabilising at 550 parts per million or below involves strong action ... but such stabilisation is feasible'. Stern's policy framework is focused on constraining the increase to 550 parts per million, at which level, he argued, 'there is around a fifty–fifty chance of keeping increases below 3 degrees [and it is] unlikely that increases would exceed four degrees'.

The link between a 60 per cent emissions reduction by 2050 and a 3-degree cap was reiterated during Stern's March 2007 visit to Australia. During an address to the National Press Club in Canberra, he said it would be 'a very good idea if all rich countries, including Australia, set themselves a target for 2050 of at least 60 per cent emissions reductions'. This would leave the planet with about 550 parts per million of carbon dioxide equivalent by 2050, and would leave us with 'roughly a fifty-fifty chance of being either side of 3 degrees above pre-industrial times'.

But according to a draft paper released in 2008 by James Hansen and seven other climate scientists, long-term climate sensitivity of 6 degrees and a doubling of pre-industrial carbon dioxide levels to 550 parts per million would produce a very different planet. Hansen reminds us that 'the last time the planet was five degrees warmer, just prior to the glaciation of Antarctica about 35 million years ago, there were no large ice sheets on the planet. Given today's ocean basins, if the ice sheets melt entirely, sea level will rise about seventy meters'. This would be the likely outcome of Stern's policy and, seemingly, also that of Australian government policy.

A number of others have followed Stern's lead. These include the former head of the Australian Bureau of Agriculture and Resource Economics, and Australia's lead delegate to the May 2007 IPCC meeting, Brian Fisher. He says the 2-degree target, with emissions peaking by 2015, 'is exceedingly unlikely to occur … global emissions are growing very strongly … On the current trajectories you would have to say plus 3 degrees is looking more likely'.

When Labor announced its 60/2050 target, it made a number of confusing and conflicting claims based on Stern's findings and various CSIRO reports. These included limiting future increases in atmospheric carbon dioxide to 550 parts per million, and setting a target range of 450–550 parts per million carbon dioxide. It also claimed to be concerned about the impacts of a 3-degree increase, and warned of 'the need for reductions in annual GHG [greenhouse-gas] emissions of 60 to 90 per cent from 1990 or 2000 levels by 2050 for countries listed under Annex 1 in the Kyoto Protocol'.

Labor also drew on the 2000 UK Royal Commission's report on Environmental Pollution, which set a cap of 550 parts per million carbon dioxide. This is odd, because in the world of climate-change science, and politics, that report is now very old—it relied on an IPCC report now 12 years out of date. Since 2000 there have been two more IPCC reports, the research has rapidly moved on, and the British government has reduced its emissions target to 450 parts per million carbon dioxide; but Labor does not refer to more recent and relevant European research. It does not mention, for example, Meinshausen's contribution to the Stern Review, which says that if greenhouse gases reach 550 parts per million carbon dioxide equivalent, there is a 63–99 per cent chance that global warming will exceed 2 degrees.

Nor did the Labor policy statement address the UK's own recognition of error, as George Monbiot noted in the *Guardian:* 'The British government has been aware that it has set the wrong target for at least four years. In 2003 the environment department found that "with an atmospheric carbon dioxide stabilisation concentration of 550 parts per million, temperatures are expected to rise by between two and five degrees". In March 2006 it admitted "a limit closer to 450 parts per million or even lower, might be more appropriate to meet a 2-degree stabilisation limit".'

What Australian Labor did was establish a target of 3 degrees and 550 parts per million, but they dressed it up as if it was aiming for something lower. Their pre-election statement says:

> In 2006, CSIRO's *Climate Change Impacts on Australia and the Benefits of Early Action to Reduce Global Greenhouse Gas Emissions* concluded that: 'Limiting future increases in atmospheric carbon dioxide to 550 parts per million, though not a panacea for global warming, would reduce 21st century global warming to an estimated 1.5–2.9

degrees, effectively avoiding the more extreme climate changes'.

This is misleading. The report referred to actually says:

> As mentioned previously, some nations view 60 per cent reductions by 2050 as consistent with placing the world on a path to achieving a 550 parts per million carbon dioxide stabilisation level. According to climate model results ... this level of mitigation would limit *21st century global warming* to 1.5–2.9 degrees, with an *additional* 0.3–0.9 degree of warming in subsequent centuries [our emphasis].

Throughout the CSIRO document, temperature increases are taken from a 1990 baseline (0.6 degrees at 1990), so that the phrase '21st century global warming to 1.5–2.9 degrees' means a total rise over pre-industrial levels of 2.1–3.5 degrees by 2100. Add in the 'additional 0.3–0.9 degree of warming in subsequent centuries', and the full temperature rise range becomes 2.4–4.4 degrees for 550 parts per million. This would clearly constitute dangerous climate change, according to anyone's measure.

This slipping and sliding of parameters, and the shift in the pragmatic goal from 2 degrees to 3 degrees, is also evident in the 2007 IPCC report on adaptation and mitigation. Of the 177 research scenarios assessed for future emissions profiles, none dealt with a target of less than 2 degrees, and only six dealt with limiting the rise to the range of 2–2.4 degrees. By contrast, 118 scenarios covered the range of 3.2–4 degrees, which suggests that the IPCC scientists, following the lead of the politicians, have also largely shifted the focus away from targets of less than 2 degrees.

The effect that carbon levels and temperature increase have on ocean algae introduces another perspective to the dialogue: what if a target of 550 parts per million were to result in the destruction of the Earth's greatest carbon sink? James Lovelock, environment scientist and proposer of the Gaia hypothesis, claims that as the ocean surface temperature warms to over 12 degrees, 'a stable layer of warm water forms on the surface that stays unmixed with the cooler, nutrient-rich waters below'. This purely physical property of ocean water, he says, 'denies nutrients to the life in the warm layer, and soon the upper sunlit ocean water becomes a desert'.

This chlorophyll-deprived, azure-blue water is currently found predominantly in the tropics, which lacks the richness of the marine life of the darker, cooler oceans. In this nutrient-deprived water, ocean life cannot prosper and, according to Lovelock, soon 'the surface layer is empty of all but a limited ... population of algae'. Algae, which constitute most of the ocean's plant life, are the world's greatest carbon sinks, devouring carbon dioxide while releasing dimethyl sulphide (DMS), which is transformed into an aerosol that contributes to greater cloud formation and, hence, affects weather patterns. The warmer seas and fewer algae that Lovelock predicts are likely to reduce cloud formation and further enhance positive climate feedbacks.

This process should be distinguished from the phenomenon of green, red, or brown algal blooms, which can occur in fresh and marine environments when phytoplankton assume very dense concentrations due to an excess of nutrients in the water. The dead organic material becomes food for bacteria, which can deprive the water of oxygen, destroying the local marine life and creating a dead zone.

Because algae thrive in ocean water below ten degrees, the algae population reduces as the climate warms. Lovelock says that severe disruption of the algae–DMS relation would signal spiralling climate change. Lovelock and Kump's modelling of climate warming and regulation published in *Nature* in 1994 supported this:

> [A]s the carbon dioxide abundance approached 500 parts per million, regulation began to fail and there was a sudden upward jump in temperature. The cause was the failure of the ocean ecosystem. As the world grew warmer, the algae were denied nutrients by the expanding warm surface of the oceans, until eventually they became extinct. As the area of ocean covered by algae grew smaller, their cooling effect diminished and the temperature surged upwards.

According to Lovelock, the end-result was a temperature rise of 8 degrees above pre-industrial levels, which would result in the planet being habitable only from Melbourne to the South Pole (going south), and from northern Europe, Asia, and Canada to the North Pole (going north).

The likelihood of dramatic effects is beginning to be recognised more widely. Stern is now saying that his 2006 Review substantially underestimated the growth rate of greenhouse gases; the impact of greenhouse gases on the levels of warming; and the degree of damage, and the risks, of climate change.

He says it also overestimated the capacity of the carbon sinks to absorb carbon dioxide, and he now warns that a 5-

degree temperature increase would, most likely, transform the physical and human geography of the planet, leading to massive human migration and large-scale conflict. Yet Stern still formally advocates the 550-parts-per-million target that he proposed in 2006, even though the data that he now uses in presentations shows that this target carries a 41 per cent chance of exceeding 5 degrees!

One possible future is that the world will fail to recognise the danger posed by a temperature rise of 3 degrees or more, and will let greenhouse-gas levels rise to 550 parts per million. In that case, it would take a long time, even under a crash emissions-reduction program, to draw down the excess carbon dioxide. As temperatures rose, driven by positive feedbacks, declining carbon sinks, and non-linear events, the climate system would have so much momentum that we would be unable, effectively, to apply the brakes at the 3-degree signpost.

In 2004, Tom Athanasiou and Paul Baer, co-founders of the climate-change social justice group EcoEquity, encapsulated the absurdity of the dilemma:

> We'd all vote to stop climate change immediately, if we only believed that doing so would be so cheap that no country or bloc of countries could effectively object. But we do not so believe. Thus we're forced to start trading away lives and species in order to advocate a 'reasonable' definition of 'dangerous' … So it's no surprise that … the advocates of precautionary temperature targets strain to soft-pedal their messages, typically by linking 2°C of warming to carbon dioxide concentration targets that can be straight-forwardly shown to actually imply a larger, and

sometimes much larger, probable warming... Climate
activists soft-pedal the truth because they think it will help,
and perhaps they are even right. Who are we to know?
Nevertheless, we also believe that the waffling is becoming
dangerous, that it threatens, if continued, to critically
undermine the coherence of our emerging understanding.
That it delays difficult, but necessary, conclusions.

# CHAPTER 12

# Planning the Alternative

What should we do once we acknowledge that 2-degree and 3-degree targets are too high, and we know that the climate system seems likely to reach more than 2 degrees of warming?

Efforts to tackle climate change, so far, have been aimed at creating a 'less bad' outcome. Society seems to be preparing simply to head into the catastrophe more slowly, which does not seem to be a very practical strategy. The alternative would be to aim for the future we want: a safe climate.

A safe climate does not involve losing the Himalayan glaciers, or endangering food production in much of China, Bangladesh, India, and Pakistan. It does not involve losing the West Antarctic ice sheet, or converting the Amazon to dry grassland. It does not involve releasing massive amounts of carbon dioxide and methane into the air by melting permafrost, or degrading nature's carbon sinks; and it does not involve having an ice-free Arctic in summer.

But to reverse or prevent these conditions is clearly a very challenging task. Perhaps it is even an impossible task.

If it had been suggested 50 years ago that humans should set out to remove the Arctic ice cap and warm the entire globe by 1–2 degrees, people would have said that this was crazy and physically impossible — that it should not and could not be done.

Fifty years on, we are well on the way to 'succeeding' in this project.

Humans now have the most powerful economy of all time. If we choose to apply this economic power to create a safe climate, and we act decisively before uncontrollable natural feedbacks are set fully in motion, we could succeed.

The first task is to understand what a safe climate means and what action is required to achieve it.

*The safe-climate zone*: When considering climate change, we can identify a range of climate conditions that are safe. Our goal must be to keep environmental conditions within that safe-climate zone, or return to that zone if conditions have, or are likely to, stray beyond it. We also need to be aware of the speed and momentum of changes in the climate system now, as it moves away from the safe zone, and later, when thinking about actions that will move it back to the safe zone.

*What is to be protected?* If climate goals are to be well formed, their underlying values need to be explicit; for example who or what are we intending to benefit? What level of risk of adverse outcomes is acceptable?

In public discussion about climate change, it is clear that motivations for action include concerns for people in various parts of the world, for other species, and for current and future generations. These concerns can be amalgamated into a concern to protect the welfare of 'all people, all species, and all generations'.

*All people*: Although the assumption in international climate negotiations is that policy is designed to benefit all people, in practice, some nations—especially the rich, high-polluting ones—plead for exceptions and special circumstances, with some success. For the international reality to shift, all the major national players must start negotiating for the global 'common good' as an extension of their self-interest. For political elites in the developed world, the motivating factors need to be both altruistic and self-interested. Compassion for people in every part of the globe must coincide with enlightened self-interest, because failure to act for all people will result in a global food crisis, mass flows of environmental refugees, and possible armed conflicts.

*All species*: In protecting other species for their own sake, we also protect ourselves, for we all belong to interdependent ecological systems. On the other hand, the elite in most countries are so reticent about acknowledging the significance of other species that their proposed climate solutions will likely take little, if any, account of the need to protect them. What we need is an ethic of cross-species compassion, together with the realisation that our own interest also requires the protection of all species.

*All generations*: In modern societies, we tend to act on the assumption that if we look after our own best interests, the future will look after itself; but this approach will not work with climate change. We need to consider the needs of future generations—both people and other species. The work of the IPCC is an interesting example. Its models and projections are run to the year 2100, a vastly longer timeline than is considered in most political processes; however, when it comes to climate change, this is not long enough. When climate impacts are suddenly found to be running 100 years ahead of schedule, as is the case in the Arctic, we are shocked, because we have to respond to events that were barely on the radar. One solution would be to follow climate models through until a stable, long-term state is reached, so that we understand the implications for all generations, rather than just for those who will live through the next 100 years. If tackled this way, we would quickly realise, for example, that the politically 'tough' target of 450 parts per million carbon dioxide would likely result in an ice-free world and a real chance of an 87-metre sea-level rise if climate sensitivity is on the high side.

Finally, of course, there is the overarching need to maintain human civilisation, in the best sense of the accumulated knowledge and wisdom that, among other benefits, informs our capacity to feed and protect the health of large numbers of people now alive.

*Risk*: As we strive to protect all people, species, and generations, we need to know how much risk people and species will be exposed to, and what humanity would consider acceptable. When approving new pharmaceuticals, and designing aircraft, bridges, and large buildings, strict risk-standards are applied: a widely used rule of thumb is to keep the risk of mortality to less than one in a million. The Apollo Program, for example, aimed to keep the risk of Saturn rockets plunging into population centres to less than one in a million. When it comes to climate change and the viability of the whole planet, it makes no sense to apply a lesser standard of risk aversion. We should aim, for example, to have less than a one-in-a-million chance of losing the Greenland and West Antarctic ice sheets, or of failing to recover the full extent of the Arctic summer sea-ice.

So far, however, governments have been accepting much higher risks in setting global-warming targets; take Stern's promotion in January 2008, for example, of a 550 parts per million carbon dioxide target which, by his own admission, means accepting devastating species loss, as well as coral reef destruction, ice-sheet disintegration, and economic damage 'on a scale similar to [that] associated with the great wars and the economic depression of the first half of the 20th century'.

Insurance Australia Group actuary Tony Coleman says insurers are familiar with managing risks to our community that are potentially catastrophic, yet, he says, when it comes to climate risk, we seem to have different parameters: '... Australia is tolerating a level of climate change risk that would be unthinkable if the nation was held to the same standards that we apply to safeguard the survival of the insurers, banks and superannuation funds that we all depend upon in our daily lives.' These levels of risk, which are less than one in 200, says Coleman, 'are completely dwarfed by the risk levels to our way of life that are now reliably attributable to potentially catastrophic climate change impacts, unless we act with urgency to rapidly reduce greenhouse emissions'.

We insist on standards of safety for individuals that are many times higher than the standards we apply to all humans and to the ecological life support systems that we, and other species, depend on.

We should not accept actions that could trigger an irreversible chain of climate-change events or produce dangerous impacts. We cannot gamble on how far we can push the system before it breaks. As is the case for civil aviation, climate-change safety policy must allow for less than a one-in-a-million chance of catastrophic failure.

*Speed of transition:* The speed of our transition to a safe-climate zone is also a critical issue. The risks posed by allowing the world to stay outside the safe-climate zone need to be assessed, along with the impacts that would be generated, ecologically and socially, by speeding up the transition. We know, for example, that species losses increase the faster that temperatures change, and we need to weigh up those sorts of impacts alongside the risks associated with a slower transition.

*Dangerous climate change versus the safe-climate zone:* How does the achievement of a safe climate relate to the more widely held goal of avoiding dangerous climate change?

The core objective of the UN Framework Convention on Climate Change, which also governs the Kyoto Protocol, is to achieve:

> stabilization of greenhouse gas concentrations in the atmosphere at a level that would prevent dangerous anthropogenic interference with the climate system. Such a level should be achieved within a time-frame sufficient to allow ecosystems to adapt naturally to climate change, to ensure that food production is not threatened and to enable economic development to proceed in a sustainable manner.

It is self-evident that the world should act to avoid 'dangerous anthropogenic [human] interference with the climate system', but making that our primary goal has created problems.

The theory of the greenhouse effect was developed more than a hundred years ago by the Swedish chemist Svante Arrhenius. In the 1960s, scientific interest in the theory began to intensify when the American scientist Charles Keeling published his findings, which showed that carbon dioxide levels in the atmosphere were systematically rising from the 315 parts per million that he had first observed in 1958. Thirty years later, scientific knowledge was strong enough to lead to the formation of the IPCC, by which time the atmospheric carbon dioxide level was 350 parts per million.

Although concern about global warming was now growing strongly, it was also clear that changing human behaviour enough to stop, or slow, greenhouse-gas emissions would require very significant changes to the economy, and that these changes would be resisted strongly. So, scientists focused their message on the need to avoid dangerous climate change — a message that proffered threats big enough to grab the attention of the political elite and, perhaps, convince them that matters other than short-term economic gain should be considered. Corporate elites, however, maintained widespread resistance to their message, so scientists opted to focus on only the strongest concerns about climate change: how could we avoid the loss of the Greenland and the West Antarctic ice sheets and the metres of sea-level rise that would occur as a result? How could widespread ecosystem collapse be avoided?

The result of this tactic was that scientists began to focus on action to avoid outright climate catastrophe. In response, the political–corporate elite set targets for change just a fraction under the levels that the scientists identified as having catastrophic consequences. Once these targets were articulated—for example, an upper-warming limit of 2 degrees, or 550 parts per million atmospheric carbon dioxide—policy inertia tended to lock them in, regardless of later changes in scientific knowledge. Scientists' concerns about identifying dangerous climate change, and about the measures necessary to avert it, were transformed into a process for avoiding catastrophe, or apocalypse, in some far-distant future. As a result, unrealistic targets have been set that, even if achieved, would see civilisation-destroying climate change.

An alternative approach would identify climate conditions that are known to be safe, and then make it the goal of public policy to get back into this safe-climate zone and avoid leaving it again. Instead of scientists being asked to identify what elevated greenhouse-gas levels might be bearable (should we stabilise at 450, 550, or 650 parts per million?), the safe-climate approach would be to ask what actions are necessary to get back to the zone in which greenhouse-gas levels are known to be safe.

*The danger of tipping points:* a major concern is the possibility that key elements of the Earth system could go through critical thresholds or tipping points (as discussed in Chapter 10) that lead to a significant increase in warming processes, such as a big jump in greenhouse-gas emissions, or to a major change that severely harms other species or human societies.

The evidence is clear that the Earth's biosphere is already in a state of dangerous climate change. Current impacts—including desertification and water shortages, extreme weather events, severe and frequent bushfires, ecological breakdown, difficulties with food production, and changes to major geophysical systems such as the Arctic—are already causing problems in many parts of the world. Pressures are building in the Earth system that will give way to even bigger changes in temperature and the environment, which means that the problems already causing concern are a relatively mild foretaste of what will come if the economy and the climate system are left to follow current trends.

While climate danger is generally cast as occurring at some time in the future, climate change is already dangerous for some people: the populations of the small nations of the Pacific, who are already abandoning their low-lying island atolls because rising sea levels and storm surges make life there impossible; people in sub-Saharan Africa badly affected by extended drought; and the Inuit people of the Canadian Arctic who can no longer move safely across the sea-ice to hunt, and whose homes are cracking and tipping as the permafrost melts.

The task, now, is to establish the boundaries of the safe-climate zone. Policy and action should be framed:

- to protect all people, all species, and all generations;
- to accept an even smaller risk of failure than the best-practice safety standards for the protection of people in civil engineering (the one-in-a-million principle) in avoiding dangerous changes to the Earth caused by climate change; and
- to keep the Earth in the safe-climate zone, rather than to simply avoid dangerous climate change.

# CHAPTER 13

# The Safe-Climate Zone

For the past 100,000 years, humans and their predecessors have survived and adapted as the Earth's temperature has fluctuated by up to 7 degrees. The current global average temperature is within 1 degree of the maximum temperature known to have occurred during the past million years, but conditions 6 degrees colder were experienced during the depths of the recurring ice ages. At a cold point 20,000 years ago, so much ice was stacked on the land that sea levels were 120 metres lower than they are now. On the other hand, 125,000 years ago, when temperatures were similar to today, the sea level was 5–6 metres higher.

The past 11,500 years since the last ice age is known as the Holocene—a period that coincides with the establishment of human civilisation. During the Holocene, temperatures have varied within a 1-degree band, although the variation has, for the most part, been considerably less. Sea levels have been almost constant over the last few thousand years of human civilisation and, more significantly, over recent centuries, when most climate-sensitive infrastructure has been built.

Coastal cities, including the special case of Venice, shipping facilities, the permanent settlement of river deltas, and other low-lying areas have survived because sea levels have moved very little.

Increasingly, however, human activity is changing the surface of the planet and also, consequently, the climate. Today we see that impact around the globe. Large parts of the land have been taken over by humans for grazing and cropping, and for cities. Wetlands have been drained on a huge scale; rivers have been regulated with dams; and forests have either been cleared, or cut into small patches by roads and clearings. As we've extracted and processed resources, and thrown away our wastes, our natural world has become very fragile, fragmented, and impacted by chemical and physical assaults.

The nation-states and the vast, fixed physical infrastructure of cities and roads that human civilisation has built will make it very hard to adapt and move across the continents if the climate were to become more changeable—if, for example, it began to swing between warm periods and glacial periods, such as those that left much of North America and northern Europe under metres, and sometimes kilometres, of ice, 20,000 years ago. While we might adapt to lower sea levels, higher seas would be catastrophic for whole cities, farming communities, nations, and coastal-wetland species.

Given our sedentary pattern of living, how can we identify a band of environmental conditions that defines a contemporary safe-climate zone? Would the relatively stable climate pattern of the Holocene and its development of agriculture and civilisations be appropriate? Can we tolerate today's temperature, which is at the top end of the Holocene range? Should we accept a summer-ice-free state in the Arctic as a normal part of the range of conditions to be included in the safe-climate zone? To maintain the Earth system's resilience, is it ecologically necessary to cycle through a summer-ice-free state periodically?

## Avoiding a summer-ice-free Arctic

During the past million years, the Arctic has been partially free of summer sea-ice for short periods, but today's circumstances are very different. In the past, this event represented the gently sloping top of the warming hill; now, however, the level of greenhouse gases, and the upward pressure on temperatures, is substantially higher. What is more, the temperature is charging through this barrier with the human foot still pressing on the emissions accelerator. The real risk is that, rather than mark the natural peak of the temperature cycle between periods of ice ages, a summer-ice-free state in the Arctic will kick the climate system into run-on warming and create an aberrant new climate state many, many degrees hotter. The last time such a warming occurred—many tens of millions of years ago—many plants and animals became extinct around the world.

An Arctic free of summer sea-ice cannot, then, be considered part of the safe-climate zone, and urgently restoring its full extent is necessary to avoid significant ecological damage and, possibly, catastrophic greenhouse heating.

In defining the safe-climate zone, it is more important to identify tangible elements of the environment that need to be restored and maintained, rather than just to focus on temperature and carbon dioxide levels.

Some features of a safe-climate policy would include:

- retaining the full summer Arctic sea-ice cover, the full extent of the Greenland and Antarctic ice sheets, and the full extent of the mountain glacier systems, including the Himalayas and the Andes;
- maintaining the ecological health and resilience of the tropical rainforests and coral reefs, with no loss of area or species;
- maintaining the health and effectiveness of the natural carbon sinks, at least, to their level of 50 years ago; and
- capping ocean acidity at a level that prevents any risk to organisms.

The appropriate temperature range and climate-system settings compatible with the maintenance of these environmental features can then be determined using the best available climate science, with a risk of loss of less than one in a million. Here are three ways of thinking about this range:

*The Hansen Arctic threshold:* In the draft paper released in April 2008, James Hansen and seven co-authors say that a carbon dioxide level of '300–325 parts per million may be needed to restore [Arctic] sea ice to its area of 25 years ago'. In other words, the amount of carbon dioxide in the atmosphere would need to be significantly reduced from the current level of 387 parts per million.

*Maintaining Arctic sea-ice thickness:* The Arctic sea-ice thinned substantially from about 3.5 metres in the 1960s to about 2.5 metres by the end of the 1980s, which was well before the beginning of the dramatic decline in ice-surface area that became apparent from the mid-1990s onwards. In the late 1980s and early 1990s, shifting wind patterns flushed much of the thick, older sea-ice out of the Arctic Ocean and into the North Atlantic, where it eventually disintegrated, replaced by a thinner layer of young ice that melted more readily in the succeeding summer. Mark Serreze from the University of Colorado says that 'this ice-flushing event could be a small-scale analogue of the sort of kick that could invoke rapid collapse, or it could have been the kick itself'. Pulses of warmer water that began entering the Arctic Ocean in the mid-1990s, which promote ice melt and discourage ice growth along the Atlantic ice margin, are 'another one of those potential kicks to the system that could evoke rapid ice decline and send the Arctic into a new state', according to Serreze. In 1989, the global average temperature was about 0.3 degrees cooler than it is currently. To restore this temperature, it would be necessary to drop carbon dioxide levels to 315 parts per million.

*An insight from the early Holocene?* For part of the period from 6000–8500 years ago, the Arctic warmed to the point that it was largely free of sea-ice each summer. A Dutch–Danish scientific team, using plant-fossil data, estimates that the carbon dioxide level during this time ranged from about 325 parts per million to a less-well-defined lower level that allowed the summer sea-ice to return.

While more research is needed before the boundaries of the safe-climate zone can be set definitively, it is reasonable and prudent to conclude from these three case studies that we should aim, initially, for at least a 0.3 degree cooling to bring the global average temperature-increase above pre-industrial levels to less than 0.5 degrees. To bring the planet within reach of this temperature, the atmospheric carbon dioxide level should be under 325 parts per million—the level that Hansen is arguing is needed to fully restore the Arctic ice.

This would also be a reasonable boundary for avoiding a range of other major climate problems, including the loss of the mountain glaciers, and the Greenland and West Antarctic ice sheets; damage to tropical rainforests; and a decline in the capacity of carbon sinks.

Hansen, discussing the impending loss of the Arctic summer sea-ice in October 2007, noted that the climate system is dominated by positive feedbacks—knock-on effects that exaggerate the current trend of the climate. These feedbacks run in both directions, so if enough of the strong, high-inertia warming feedbacks were stalled, or turned around, and the Earth was cooled for a while, the climate system would then run in the opposite direction. If humans decided to initiate a sufficient cooling, natural feedbacks would complete the job.

## Cooling the Earth

The Earth is already too hot, and there's already too much carbon dioxide and other greenhouse gases in the atmosphere. The first key step to fix this is to stop adding to the heating processes—greenhouse-gas emissions need to be cut to zero. The second step is to remove from the air the excess carbon dioxide that is keeping the planet too hot. The third step, because time is short and there is already so much heat in the system, may be for humans to cool the Earth directly.

To cut greenhouse-gas emissions, we will need to reduce those warming agents that have a short life. Methane, for example, has a relatively short life in the atmosphere of about a decade, so cutting methane emissions would have an effect relatively quickly. Measures to achieve this include stopping coal, oil, and gas mining (to stop methane leakage); re-engineering waste disposal (trapping methane as an energy source); changing irrigation methods and varieties of rice cultivation; and decreasing the commercial herding of ruminant animals, especially cattle.

We must also stop emitting greenhouse gases, including carbon dioxide, and heating agents, such as black soot, urgently. This is essential, because carbon dioxide is acidifying the upper ocean, preventing marine organisms from forming calcified shells and exoskeletons. If this continues it will lead to major marine animal and plant extinctions in the not-too-distant future. Black soot is a short-lived warming agent that is washed out of the air by rain in a matter of days; cutting its emissions would have an immediate effect. By dirtying ice, black soot also accelerates glacier and ice-sheet melting—particularly in the Himalayas, because one-third of black-carbon emissions come from India and China. Programs to cut black-soot emissions—for example, by ending the use of coal for heating, stopping diesel use, and by providing energy-efficient and smoke-free cookers to rural communities across Asia—would have an immediate and dramatic effect in reducing the heating effect.

*Zero-carbon Britain: an alternative energy strategy*, published in 2007, is one of many research reports that demonstrate the feasibility of building a post-carbon economy. Many of the practical technologies and solutions are also surveyed in Chapter 20 of this book.

We must also remove excess carbon from the air. We cannot return to a safe climate if we only *cut* emissions to zero, because carbon dioxide remains in the atmosphere for so long. Estimates by Matthews and Caldeira from Stanford University indicate that around 200 billion tonnes of excess carbon needs to be drawn out of the atmosphere to achieve the 0.3-degree decrease in the global temperature that is necessary.

Techniques for trapping carbon that is already in the atmosphere include boosting the natural terrestrial processes (re-afforestation); and producing agricultural charcoal, known as bio-char, which is sequestered in the ground.

Such large-scale, relatively low environmental-impact methods depend on growing plants that naturally absorb carbon dioxide. Growing these in the extremely large quantities necessary to draw down substantial amounts of carbon, however, may conflict with land use for nature conservation or food production. This, along with issues such as water availability and social impacts, needs to be considered in planning such schemes.

Although it is necessary to reduce human greenhouse-gas emissions to zero as quickly as possible, there is a critical side effect. Most carbon dioxide generated by human society is produced by deliberately burning fuels such as coal, oil, gas, and wood, or by unintentionally burning plant material in bushfires, but these processes also produce aerosols, including smoke, and small-particle pollution such as soot, dust, and sulphate particles.

If we were to stop burning fossil fuels tomorrow, the aerosols that cool our planet would be rained out of the air in about ten days. Without these aerosols, which mask roughly half the heating effect caused by carbon dioxide, there would be a sudden jump in temperature. Stopping all carbon dioxide emissions could produce a short-term warming of one-half to one degree. Cutting black carbon-soot emissions would offset some of this effect.

Removing aerosols causes steeper warming the more quickly that fossil-fuel combustion is cut. If fossil-fuel combustion were to be cut to zero in two decades then, assuming a mid-range climate sensitivity, the loss of the related aerosols plus the warming already in the system could produce a warming of more than one degree in twenty years. This would be highly destructive to our ecosystems.

Reducing fossil-fuel combustion to zero in 50 years will also produce a rate of warming far beyond the capacity of most ecosystems to cope, because of the aerosol cooling lost. Cutting fossil-fuel combustion much more slowly, to zero in a hundred years, would have the same effect because, particularly in the latter part of the timeline, more carbon would be kicking into the atmosphere from failing natural carbon sinks, exacerbating the long-term trend.

Slowing the rate of reduction of fossil-fuel combustion may then make the warming problem from aerosol reduction less severe in the short term, but worse in the long-term.

These 'damned if you do, and damned if you don't' problems are known in the fields of science, politics, and economics as 'wicked problems', a concept first articulated by design and planning theorist Horst Rittel. A 'wicked problem' describes a complex set of interrelated and circular problems which are resistant to resolution and where any solution is not good or bad, but only better or worse. Each 'wicked problem' is unique and, effectively, offers only one chance to achieve the least-worst resolution, because poorly constructed 'solutions' can compound the problem and, in many cases, there is a limited time horizon for effective action.

Getting to the safe-climate zone will take time. But, as each year slips by, the impact of warming and the problem of positive feedbacks takes us further away from that zone.

How long will it take the Earth to cool sufficiently if we achieve zero greenhouse-gas emissions and start reducing atmospheric carbon dioxide? This will depend, in part, on how long it takes us to make the major economic change necessary to achieve zero emissions and to put in place a system to capture and sequester excess carbon dioxide from the atmosphere. Historical precedents for rapid industrial change, such as the Asian 'tiger economies' and the information-technology revolution, suggest that it could be achieved in two to three decades. The economic restructuring achieved during World War II shows that a fast economic transition like this is possible.

However, even after major economic changes were made, drawing down excess carbon dioxide from the air could take 50 to 100 years (even at a high depletion-rate of 6 billion tonnes a year), because of the carbon still being emitted, and the time necessary to develop, build, and maintain such large-scale processes.

While Arctic cooling would start before all the excess carbon dioxide was taken out of the air, it could still be a century before it returned to a safe-climate condition. Over that lengthy amount of time, the Earth would still be subject to rising temperatures. A 1-to-2-degree rise above the present temperature is not out of the question, even if we established a zero-emissions economy in two to three decades — especially when the aerosol dilemma is taken into account.

The risk is real that during the early decades of the transition, major damage could be done to Earth's ecosystems, such as tropical rainforests. It is also possible that the permafrost, and other sources of natural carbon, could be so strongly mobilised, and the natural carbon sinks so damaged, that the process of taking carbon dioxide out of the air would be overwhelmed. In one century, enough ice could be lost from the Greenland and West Antarctic ice-sheets to raise sea levels by several metres. If most of the ice in the Himalayas were lost, food production in nations from the Indian sub-continent to China would be drastically reduced. These would be civilisation-disrupting changes, even if run-on heating was avoided.

The grim reality is that the Earth is too hot right now. Even a zero-emissions strategy and a monumental effort to pull excess carbon dioxide out of the atmosphere will not achieve the necessary cooling soon enough. We must consider the third strategy — cooling the Earth directly.

Natural ecosystems play a role in moderating the Earth's temperature. Marine plankton releases a gas, dimethyl sulphide, that disperses into the air and helps to form dense, sunlight-reflecting cloud; there is also some evidence that water evaporating from forests carries bacteria into the atmosphere to aid the formation of light-reflecting clouds. So boosting the health of marine plankton could help, as could re-establishing forests on a large scale; however, as global warming grips the planet, it is becoming more difficult to maintain biological systems, let alone to re-establish large areas of forest. If they are to cool the atmosphere, the Earth's ecosystems may need even more help.

In 1992, a US National Academy of Science report on greenhouse warming discussed climate geo-engineering, including very large-scale projects that would deflect a small proportion of the solar radiation striking the upper atmosphere, in order to produce a small cooling. The possibilities considered ranged from the science-fiction-like placement of reflectors in space, to copying the cooling effect of volcanic eruptions by pumping sulphates, or other particles, into the upper atmosphere, where they would last for a year or two.

More recently, scientists such as Ken Caldeira and Nobel-laureate Paul Crutzen have studied proposals to pump sulphate aerosols into the stratosphere, and it seems that this measure could fully cancel the warming caused by greenhouse gases and other warming agents, such as black soot. The sulphate program could be implemented within a few years and have an immediate cooling effect. However, it is not a substitute for a zero-emissions program, or an excuse to continue emitting greenhouse gas. Nor is it a safe solution in the long-term, since any premature end, or interruption, to such a program—through war or economic recession, for example—would subject the globe to a major heating-pulse within two years.

The greatest value from a sulphate geo-engineering program, with the least ecological risk, would be to carry it out as soon as possible, for as short a time as possible. Every effort should be made to reduce the necessary intensity of any geo-engineering program—for example, by bringing down methane and black-soot pollution as fast as possible.

Humans have been unintentionally geo-engineering the Earth for a long time. The first major intervention was the use of fire to reshape ecosystems over large areas; the second was the introduction of large-scale land clearing for farming and other purposes; and the most recent has been industrialisation, which moved massive amounts of materials into the air, the waters, and the land. The warming effects of this pollution are now, belatedly, recognised as very dangerous; but it looks as though simply stopping this geo-engineering, without first returning the Earth to a viable state, is not a workable option.

It is critical that temporary atmospheric geo-engineering should complement, rather than replace, a zero-emissions industrial structure that would remove the full excess carbon dioxide from the air as fast as possible. If it is not part of such a package, atmospheric geo-engineering should be rejected outright.

If this kind of program seems desperate, it is only an indication of the desperate straits our planet is now in. We must consider the least-worst options to save the Earth. These are absolutely necessary to stop the climate becoming so warm that a return to the safe zone is beyond reach.

CHAPTER 14

# Putting the Plan Together

This book started in the Arctic, because shrinking sea-ice is one of the triggers which shows us that climate change is already dangerous. To recover the Arctic's climate, the goal developed in Chapter 13 is to replace global *warming* with global *cooling* sufficient to drop the temperature enough to allow the full return of the summer sea-ice. This is the only way we can avoid the domino effect of sea-ice loss, the albedo flip, a warmer Arctic, a disintegrating Greenland ice sheet, more melting permafrost, and the whole catastrophe.

If we fail to stop global warming, life in the future will be hellish—not because of what we will do from now on, but because of we have already done. Based on a conservative view of climate sensitivity:

- Human emissions, so far, have produced a global warming of 0.8 degrees.
- Our non-carbon dioxide greenhouse-gas emissions are adding about another 0.7 degrees to potential global warming. This amount of heating is offset by

aerosols, which have a temporary cooling effect of approximately 0.5–1 degree, although this figure may well be too low.

- There is more warming to come as a result of 'thermal inertia', which refers to the delayed temperature effects produced by rising carbon dioxide levels. Only about half of the temperature increases will appear within 25 years, another quarter will take 150 years, and the last quarter may take many centuries to show up. This is because the oceans are continuing to absorb much of the heat accumulated as a result of rising carbon dioxide levels. Once the oceans cannot absorb any more, the heat will build up in the atmosphere. Thermal inertia and other lags in the system will take the total long-term global warming induced by human emissions, so far, to 1.4 degrees (although some warming will take a very long time to manifest, and will be affected by the extent to which the carbon cycle, over time, draws down the atmospheric concentration).

- The loss of the Arctic ice would also produce an increase of approximately 0.3 degrees, due to the albedo flip, although it will take time for this to happen.

- If total human emissions continue at their present level for two more decades, this is likely to add at least 0.5 degrees to the system by 2030; however, on current trends, emissions are projected to increase 60 per cent above present levels by 2030.

If we keep acting as we have, the Earth's atmospheric temperature will very likely be at least 2 degrees warmer by mid-century, with more warming to come.

A world free of the imminent threat of climate catastrophe would be one in which the Arctic basin again gleams white with sea-ice, and in which human ingenuity and determination is sufficient to cool the Earth back to the safe-climate zone. Although some people are incredulous when they first hear this proposition, we are yet to see a reputable climate researcher state that the Arctic could remain free of sea-ice long term without dangerous climate change occurring.

We can cool the planet, get back the Arctic sea-ice, and preserve the great polar and high-mountain ice-sheets — or watch the system spin out of control. There is no middle way between these stark options. It is not a matter of how much more greenhouse gas we can add to the atmosphere; it is a matter of what means we must use, what speed we must attain, and what extent we must reach, as we take action to draw down the current levels of greenhouse gases to a safe level, in time to avoid catastrophic climate change.

Any proposal for a goal of higher than 0.5 degrees warming would be foolhardy. The only alternative conclusion (which we do not support) is that it is safe to leave the Arctic sea-ice melted, and to plan for a long-term target much higher than the current warming of 0.8 degrees, on the assumption that such a course of action would not involve the planet crossing dangerous climate tipping points.

This second approach is, implicitly, the view of all the major nations and players involved in setting climate policy. Their challenge is to provide a reasoned argument explaining why it is a safe course of action to leave the Arctic Ocean free of ice in summer. We are not aware of any evidence that would support such a proposition.

NASA's James Hansen told scientists and others at the American Geophysical Union conference in San Francisco in December 2007 that we, as a species, passed climate tipping points for major ice-sheet and species loss when we exceeded 300–350 parts per million carbon dioxide in the atmosphere. He said that this point was passed decades ago, and that climate zones such as the tropics, and temperate regions, will continue to shift, and the oceans will become more acidic, endangering much marine life. He added: 'We either begin to roll back not only the emissions [of carbon dioxide] *but also the absolute amount in the atmosphere,* or else we're going to get big impacts ... We should set a target of carbon dioxide that's low enough to avoid the point of no return [our emphasis].' Hansen estimated that target to be 300–350 parts per million carbon dioxide, concluding: 'We have to figure out how to live without fossil fuels someday. Why not sooner?'

To restore the Arctic sea-ice, James Hansen and his co-authors have explicitly identified the target as being in the range of 300–325 parts per million carbon dioxide. This is consistent with work by Hansen, before the Arctic summer of 2007, which pointed to the need for a cap that was a safe amount less than 1.7 degrees:

> Earth's positive energy imbalance is now continuous, relentless and growing ... this warming has brought us to the precipice of a great 'tipping point'. If we go over the edge, it will be a transition to 'a different planet', an environment far outside the range that has been experienced by humanity. There will be no return within the lifetime of any generation that can be imagined, and the trip will exterminate a large fraction of species on the planet.

More recently, in court testimony in Iowa, Hansen reaffirmed his view: 'I am not recommending that the world should aim for additional global warming of one degree. Indeed, because of potential sea level rise, as well as the other critical metrics ... I infer that *it is desirable to avoid any further global warming* [our emphasis].'

## Global equity

Until recently, most players in the climate-policy arena assumed that while global-warming emissions needed to be cut substantially, they did not need to be reduced to zero, so it would be fair for all people across the globe to share a reduced annual greenhouse-gas limit. Poor people could keep increasing their fossil-fuel use until their emissions reached the limit, and people in rich countries would need to keep reducing their emissions until they reached the same per capita level (a principle known as 'contraction and convergence').

But it is now clear that greenhouse-gas emissions must be cut to zero, levels of carbon dioxide must be drawn down and, most likely for some decades, the planet must be actively cooled. What, then, is a fair way to share these global tasks?

Our proposed safe-climate strategy is based on the protection of 'all people, all species, all generations'; but people and nations have contributed very unevenly to global warming. The developed economies are responsible for most of the historic atmospheric carbon emissions (and most emissions since 1990), and they have the responsibility and the capacity to provide resources to the world's poorer nations to create a path to development that preserves a safe climate. In a September 2007 report, the global investment bank Lehman Brothers called for a 'global warming superfund', and strongly implied that nations should pay into it on the basis of their historical emissions.

More systematically, a Greenhouse Development Rights framework has been designed by the US-based climate think-tank EcoEquity to support an emergency climate-stabilisation program while, at the same time, preserving the right of all people to reach a dignified level of sustainable human development that is free of the privations of poverty. The framework quantifies national responsibility and national capacity. Its goal is to provide a coherent, principle-based way of thinking about the national obligations to pay for emissions-reduction, and to adapt to changing climates.

There is no significant benefit to any country in continuing with a high-emissions economy, and *all* new investments should fit the modern zero-emissions paradigm. But what about cutting emissions from existing plant and equipment? The cost of this task, and of providing an adequate and secure energy-supply for all people in all countries, should be borne on the basis of past responsibility and present capacity.

There is no country, or class of people, rich or poor, that will benefit in any durable way from a greenhouse-devastated world. All people, in every country, face the need for change. There is no doubt, however, that the lion's share of the problem was caused by the rich countries and classes—initially unknowingly, but for at least the last two decades, wilfully—and so ethical equity principles would lay the bulk of the cost at their feet.

### The challenge

In light of the strong goals we have proposed—the need to develop large-scale drawdowns of carbon dioxide, and the need to solve the aerosols dilemma through geo-engineering or other means—there is a legitimate concern about whether, in present social and political circumstances, such deep and rapid change is possible. Very large levels of investment would be required to solve problems, develop and implement new technologies and solutions, and restructure the economy rapidly. It is hard to imagine that the unity of purpose required for such a transition could be attained in normal political and social circumstances.

Fortunately, human societies have another mode that they turn to in times of great need: the state of emergency. The form of emergency required to tackle the climate crisis will be different, in important ways, from more familiar emergencies: it will require coordinated global actions, it will be a long emergency, and the world will be very different, in many ways, when it is over.

It is possible that governments will grasp the gravity of our situation, recognise the emergency, and create the social, administrative, and economic circumstances required to deal with the climate crisis. More likely, we will need to create a popular movement and a deliberate advocacy program to create the necessary political will. Failure to declare a state of emergency is likely to result in a profoundly ineffective response to the climate crisis.

The risk today is that we continue to treat the climate crisis as something that lies in the future, and we continue to talk about reaching climate targets only years, or even decades, hence. So far, when practical difficulties arise with targets, we have re-calibrated the future, deluding ourselves that more warming is reasonable. It is not. Will we continue to recalibrate the future as the truth of climate change becomes increasingly inconvenient? That path will condemn our descendants to accept the bitter truth that we allowed two degrees of warming to become three, and then four, as the seas engulfed their cities and farmlands.

The alternative is to move beyond politics and business as usual, and into emergency mode.

PART THREE

# The Climate Emergency

'The era of procrastination, of half-measures, of soothing and baffling expedients, of delays, is coming to a close. In its place we are entering a period of consequences.'

– Winston Churchill, November 1936

# CHAPTER 15

# This Is An Emergency

On 13 April 1970, some 321,000 kilometres from Earth, the Apollo 13 spacecraft was hit by an explosion that resulted in a loss of oxygen, potable water, and most electrical power. The access panel covering the oxygen tanks and fuel cells, which extended the entire length of the main craft's body, had been blown off. Apollo commander Jim Lovell's laconic message, 'Houston, we have a problem,' signalled a technological failure so great that mission objectives were abandoned. The moon landing was aborted.

The priority of the astronauts onboard the craft was survival at any cost. Life-support systems were at risk, and energy use had to be cut to a minimum, since little power was available. The crew shifted to their tiny lunar module — an emergency procedure than had been simulated during training — and abandoned the main craft, to which the module remained attached. But the lunar module was equipped only to sustain two people for two days; now, with insufficient capacity to keep the air clean or to heat the module to a habitable temperature, it needed to sustain three people for four days. There was no precedent, no manual, and no set of pre-tested solutions; but there was a driving imperative that was reinforced by mission control in Houston: 'Failure is not an option!'

A sequestration filter was invented on the run while carbon dioxide rose to dangerous levels. With inadequate mechanical control, the astronauts had to negotiate course alterations while engineers on the ground calculated the best way to use auxiliary motors to position the craft for the return journey.

Under this level of pressure, the on-the-run problem-solving required ingenuity and intense teamwork. The outcome was in doubt up to the last moment, but the crew made it and survived. The mission was deemed 'a successful failure'. Careful planning and training (including allowing for the possibility of having to jettison the main craft), strong cooperation between all involved, creative off-the-wall solutions, and a great measure of good fortune had combined to save the day.

Today, Earth faces a similar degree of peril, and its message can only be: 'People of the world, we have a problem.' Our planet's health and its capacity to function for the journey through time are now deeply imperilled. We stand on the brink of climate catastrophe.

Like Apollo 13, we have only one option: to abandon our life-as-normal project, hit the emergency button, and plan with all our ingenuity how to survive and build a path for a return to a safe-climate Earth. We have to act with great speed, determination, and ingenuity. Our life-support systems—food, water, and stable temperatures—are at risk, and our consumption of fossil fuels is unsustainable. Energy use must be cut. The voyage will be perilous, and will require intense and innovative teamwork to find and mobilise technological and social answers to as-yet-unidentified problems. Putting aside mantras about high costs, our collective actions need to be driven, instead, by the imperative: 'Failure is not an option!' If we do not succeed, we will lose most of the life on this planet.

Lacking its main motors and with uncertain technological control functions, Apollo 13 had only one chance to position itself in exactly the right trajectory so that the moon's gravitational force would pull it back to Earth safely. We, too, have only one chance to get global warming under control and to guide the planet back to the safe-climate zone. If we do the wrong things, or we set our approach incorrectly and don't do enough, there will be no time for a second chance.

We have already entered an era of dangerous climate change. If left unchecked, the dynamics and inertia of our social and economic systems will sweep us on to ever more dangerous change and then, most likely within a decade, to an era of catastrophic climate change.

If the response to global warming continues to be contained within the current all-too-narrow parameters, it will guarantee disaster. Given the lessons from the Arctic summer of 2007 — let alone all the other early-earning signs that climate scientists are noting increasingly — allowing warming to reach even 2 degrees, let alone the increasingly advocated 3 degrees, is reckless.

This is our emergency.

# CHAPTER 16

# A Systemic Breakdown

The Apollo 13 emergency put just three lives at risk. Large emergencies triggered by flood, fire, famine, earthquake, or disease may affect millions. Across such diverse circumstances, the usual approach is to direct all available resources towards resolving the immediate crisis and to relegate non-essential concerns to the back burner.

Today, there is a practical argument that we should focus all our attention exclusively on the climate crisis, because it will take a huge effort to solve; but we need to ask whether there are other issues that will be seen, in retrospect, to have caused major problems if they were to be ignored, either because they are of great moral significance or because they seem more compelling in the short term.

The unambiguous answer is that there are several key concerns that must be resolved together with the climate crisis. There are those—such as peak oil, severe economic downturn, warfare, and pandemics—that cannot be ignored because their impacts on all people are so great. There are also ethical problems we should not ignore, such as poverty (including the adequacy of food supply at an affordable price) and the need for biodiversity protection.

The intertwined problem of climate and dwindling oil reserves is a good example. The continuing discovery of geological reserves of cheap conventional oil cannot keep pace with growing world demand, and the crisis point for oil production and consumption, commonly referred to as 'peak oil', is a reality. Its emergence is reflected, in part, in rising oil prices and in the expectation that they will continue to increase as the gap between supply and demand increases in coming years. In Australia, the 2007 Queensland state government's task-force report *Queensland's Vulnerability to Rising Oil Prices* found 'overwhelming evidence' that world oil production would reach an absolute peak in the next ten years; at the same time, the US Department of Energy predicts that demand for this declining resource will have increased by 24 per cent by 2020.

Clearly, we cannot resolve the global-warming threat before peak oil demands our attention in a very practical way. Nor can we delay resolving the climate issue: the restructuring needed to solve the peak-oil problem will take at least ten to 20 years, yet the climate solution demands major economic changes in the same time-period. The two problems must be considered together with integrated solutions.

At the same time, we must find appropriate solutions that also address other challenges. To achieve a safe climate and eliminate human greenhouse-gas emissions, we need to apply many resources simultaneously; we need to take large amounts of excess carbon dioxide out of the air, and we need to actively cool the Earth; at the same time, we need to maintain adequate supplies of affordable food, and secure survival of the world's biodiversity.

The production of fuel substitutes as a solution to peak oil is an example of the sort of challenge we are encountering in trying to achieve integrated solutions. Faced with increasing dependence on imported oil (which is continually rising in price), the US government, among others, is actively encouraging the diversion of food crops to the production of ethanol, a petrol substitute. Together with climate-related food-production problems, the ethanol 'solution' has contributed to global food shortages and sharply rising prices, and has particularly affected the poor and malnourished.

For transport, the alternative to focusing on the replacement of one organic fuel source (petrol) with another (ethanol) is to actively reduce the demand for energy — for example, by replacing current cars with vehicles designed for ultra-efficiency (such as electric vehicles charged from renewable sources), or by creating infrastructure that enables people to switch from car travel to public transport, walking, or cycling. The need for mobility could also be reduced by improving urban planning, or by making use of electronic 'virtual travel' such as video-conferencing. So far, these demand-reduction strategies have not been given political prominence.

The connection between climate, rising oil and food prices, the financial crisis, and economic recession is another example of the interplay between critical issues. Since the 1987 Wall Street crash, monetary authorities have used credit expansion — and condoned the development of a whole new range of dubious financial services and practices — as a tool to promote consumption and to stop the economy spiralling into recession. But now strong inflationary pressures are being driven by rising oil and food prices, and by expansionary war expenditure for Iraq and Afghanistan—itself motivated in part by oil—and these are being financed by large deficits. The consequence has been financial crisis, but monetary authorities are now not as free to use credit expansion to increase demand, and this slowdown may have its own negative impacts on climate: if a recession is allowed to run its course, there could be less money available to invest in responses to the climate and peak-oil crises. On the other hand, if governments invest in traditional public infrastructure areas in order to 'prime the economic pump', the result may be more roads and freeways, which will exacerbate climate and peak-oil problems. Only if pump-priming investment is framed with the climate and peak-oil problems in mind will the response to recession produce a beneficial cycle of change.

A systemic crisis often arises when many problems come to bear on one key issue. In the late 1960s and early 1970s, there was concern about the likelihood of future large-scale food shortages, because population was growing rapidly and there was a fear that food production would fail to keep pace. Overall, populations did not continue to grow as fast as expected, and food supply expanded more rapidly—a result of the 'green revolution', which utilised new strains of higher-yield crops and increased inputs of water and fertiliser. This worked in the short term, but required more water per unit of agricultural output, and increased the use of nitrogen-based fertilisers, which release global-warming nitrogen oxides.

Now, a whole series of problems are driving a wedge between potential demand and actual supply for water. In many areas, and especially in the heavily populated developing countries of Asia, extra water was made available through tube wells, or bores. This worked for decades, but groundwater stocks are now running out, and in some places the water is naturally contaminated with arsenic, causing serious health problems. Global warming has also affected rainfall, so that there is less usable water available at the same time as urban demand for water is increasing.

An increase in unpredictable weather events and changing climates — for example, unusual monsoon patterns — are making it harder to maintain food supplies. Most available high-value land with agricultural potential has been utilised, so there is, increasingly, less opportunity to expand agricultural areas and to replace land that has been damaged by erosion and salinity. The UN's 2007 report *Global Environment Outlook: environment for development* found that total arable land has reached a plateau at 14 million square kilometres, while the area under cereal crops dropped from 7.2 to 6.6 million square kilometres between 1982 and 2002.

The continuing growth of the global population — and of incomes for many people in industrialising countries — is increasing demand for food, just as food crops are being diverted to biofuel production. As a result, the prices of many food staples have been rising sharply. Between 1974 and 2005, world food prices fell by three-quarters, in real terms. Since then, that trend has reversed, with the price of wheat almost doubling in 2007, and the prices of maize, milk, and oilseeds reaching near-record highs. In 2007, the food-price index, published by *The Economist*, increased by 75 per cent.

It seems that the price of food — or the supply of affordable food — is becoming a key indicator of a new phenomenon: a multi-issue crisis of sustainability that incorporates food, water, peak oil, and global warming.

At the same time, the natural physical infrastructure on which all living things depend is being put under more and more stress. Marine ecosystems are increasingly breaking down due to over-fishing, while forests in many parts of the world are being cleared on a vast scale. Where they're not fully cleared, forests often are being broken up into isolated islands that have a greater chance of being invaded by pest species and less capacity for native species to move between areas in response to fire and drought.

An overwhelming case has been put forward which says that we should not focus on climate change exclusively. If we ignore the many issues that could undermine life and wellbeing, we may, if we are lucky, solve the climate crisis only to find we have crashed the planet's life-support systems in some other way.

Since the beginning of the industrial revolution, we have failed to build and maintain a system that has enabled modern society to ensure its own sustainability and that of other living species. Now, we have a sustainability crisis with a multitude of serious symptoms. An effective governance system would anticipate and prevent threats to sustainability, and would also have the capacity to restore the Earth and society to its safe zone as soon as possible.

# CHAPTER 17

# When 'Reasonable' Is Not Enough

In November 2007, UN secretary-general Ban Ki-Moon told the world that global warming is an emergency, and 'for emergency situations we need emergency action'. Why, then, has climate politics moved in such a painfully slow manner? How can the impasse be resolved between urgent action, based on the science, and action that seems 'reasonable' in the current political environment?

It seems as if there are two great tectonic plates — scientific necessity and political pragmatism — that meet, very uneasily, at a fault line. Some examples may help to illustrate the tensions and compromises that result from trying to balance the two factors:

- In 1996, the European Union's Environment Council ignored advice from the advisory group on greenhouse of the World Meteorological Organisation, the International Council for Science, and the UN Environment Programme that an increase in the global average temperature of greater than 1 degree above pre-industrial levels 'may elicit rapid, unpredictable and non-linear responses that could lead to *extensive ecosystem damage*' [emphasis added]. Instead, they advocated a 2-degree cap, even though that figure was described as an upper limit 'beyond which the risks of grave damage to ecosystems, and of non-linear responses, are expected to increase rapidly'.

- The Stern Review identified a need, based on its reading of the science, for a 450-parts-per-million (or 2-degree) cap of carbon dioxide levels, but then said that this would be too difficult to achieve and advocated a 550-parts-per-million (or 3-degree) cap instead.

- In 2007, under Kevin Rudd, the Australian Labor Party's pre-election climate-policy statement effectively supported a 3-degree cap, despite data quoted in the statement itself that unequivocally demanded a much lower target.

- The IPCC has not called for climate modelling stabilising temperatures at less than 2 degrees, despite the evidence that the safe zone is much lower. Although the IPCC says its role is to simply represent the science, not to advocate policy, this seems to be a case of the IPCC allowing political norms to limit the scope of the research that it encourages or reports.

- Many climate and policy researchers, while privately expressing the view that the 2-degree cap is too high for a safe-climate world, have, nevertheless, publicly advocated less effective goals, because they perceive them to be more acceptable. Their argument is that they 'wouldn't be listened to' if they said what they really thought.

- Some climate-action advocates speak of the need to occupy the 'middle ground', or to be at least 'heading in the right direction', because 'it is always possible to go further later on'. This stance turns risk-aversion on its head by failing to consider worst-possible outcomes. At the same time, it is politically advantageous because it obviates the need to talk about preventative actions that are currently perceived to be 'extreme'. As a result, much advocacy aims for a direction-setting minimum requirement, rather than for a clear statement of what is needed.

- During 2007, the position of the Australian Conservation Foundation was that emissions should be cut '60 to 90 per cent' by 2050 (a 60 per cent cut would leave emissions in 2050 at *four times* the level required of a 90 per cent reduction). Yet, in a 2008 preliminary report, economist Ross Garnaut told the new government that a 90 per cent cut may be necessary, and that 60 per cent was far from enough.

In all these examples, we see a reluctance on the part of organisations and people to go beyond the bounds of perceived acceptability. The result is advocacy of solutions that, even if fully implemented, would not solve the problem. We have a sense that many of the climate-policy professionals—in government, research, community organisations, and advocacy—have established boundaries around their public discourse that are guided by a primary concern for 'reasonableness', rather than by a concern for achieving environmental and social sustainability.

Many people whose work centres on climate change have been struggling for so long to gain recognition for the problem—having had to cope with a lack of awareness, conservatism, and climate deniers—that they now have deeply ingrained habits of self-censorship. They are concerned to avoid being dismissed and marginalised as 'alarmist' and 'crazy'. Now that the science is showing the situation to be far worse than most scientists expected only a short while ago, this ingrained reticence is adding to the problem.

A pragmatic interdependency links many of these players in a cycle of low expectations and poor outcomes. Here is an outline of the concerns of some of these key players, based on conversations and correspondence we have had with them. The cycle is a merry-go-round, so it matters little where it starts.

Under pressure to stick to the science and avoid expressing an opinion, a climate scientist may take the view that society needs to make the judgement about what it determines to be dangerous climate change: 'It's not for me, as a scientist, to tell you what's dangerous or what the political target ought to be. I try to inform the debate by explaining what the risks actually are at these various levels, and by offering policy options that society could consider.'

Community-based climate-action groups, often lacking detailed technical knowledge, will respond by saying that they are not about to doubt the views put forward by the science professionals, which they hear from the media and from the IPCC: 'We have to trust in their abilities to lead us. They are the ones who know—we can't say things that they haven't, and we can't speculate on what a few scientists might be saying, if it isn't in the IPCC reports.'

Large climate-group and environment managers often join the conversation, suggesting that they agree with strong goals and urgent action, but they are worried that if they promote them, their lobbying wouldn't be taken seriously: 'It is more important to agree and campaign on targets that are heading in the right direction, than that we have discussions about what the targets should be. It is always possible to go further, or call for more, later on.'

The consequence is that even those politicians who are climate friendly feel constrained: 'I can't go further than the environment movement. I'd look extreme if I did.' And: 'I know our party's position will have to be strengthened because the science has changed, but that can't happen until after the next election. Our policy is now set. I wish we could go further, but some people are worried that I will look too extreme in the electorate.'

Deep inside public administration, where climate policy is processed, there is an avoidance of the political: 'Although our climate-science manager agrees with your targets ... she has to stick to using scientists, not lobbyists, and science, not policy. She needs to be persuaded that setting targets and trajectories is fundamentally a climate-science issue, not a political one. If, on the other hand, we can find a scientist to make the case for real targets that you have made, this would help a lot, but the scientists say that target-setting is political, and outside their terrain.'

Businesses, meanwhile, remain constrained by their commercial interests: 'You might well be right that 60 per cent by 2050 is not enough, but the people I talk to wouldn't believe anything tougher. Our business is one of the good ones — we know that this is a big problem, but if we are going to engage the wider business community, we can only go so far.'

It seems that everyone is waiting for someone else to break the cycle. But how can this be done? Part of the problem seems to be fear: those who might become the first to move to a tougher position are worried about becoming isolated or losing credibility. This could be overcome if a broad range of players agreed to move together. Another approach would be to start with the question, 'What do we need to do to achieve a safe-climate future?' rather than, 'How far should we move from our existing position?' To the best of our knowledge, no advocacy group, government, or political party in the United Kingdom, the United States, or Australia has ever asked scientific researchers to prepare a safe-climate scenario, or a 'what if' plan. Such a plan would allow various climate advocacy activists and lobbyists to get a tangible feel for what needs to be done, before having to commit to a specific plan, or advocating challenging new goals for action. Once safe-climate scenarios have been developed and supported by a range of the leading climate-policy players, they can be taken to a wider audience for discussion. In this way, we can overcome the 'credibility' blockage. Without a doubt, people and organisations that are sufficiently confident in their views to raise difficult questions can more easily explore the 'what if' strategy.

Reticence on the part of advocates to push for serious action also stems from the pervasive view in politics that everything is subject to compromise, and that trade-offs are the norm: argue less for what you really want than for what seems 'reasonable' in the give-and-take of normal political society. And when some brash advocates do argue for what really needs to be done, it is simply assumed that they are making an ambit claim: an initial demand put forward in the expectation that the negotiations will prompt a lesser counter-offer and will end in compromise.

While this mindset is widespread, there are domains from which it has been banished. When it comes to public safety, society knows that compromise and negotiable trade-offs must not apply: bridges, buildings, planes, large machines, and the like must be built to risk-averse, high standards, which are applied rigorously. If standards are not met and structures fail, corporations, governments, and regulatory bodies are held to account. We have learned from trial and error that a 'no major trade-off' policy in public safety is necessary to avoid death or injury to our citizens.

With global warming, however, we do not have the luxury of learning by trial and error. We have left the climate problem unattended for so long that we now have just one chance to get things right by applying a 'no major trade-off' approach without a trial run. It will be a particular challenge for decision-makers who have grown up in a political culture of compromise.

Because the last emergency mobilisation on this scale was during World War II, few people today have any direct experience of a situation like this; however, there is plenty of history from which to learn, and expertise available, to plan for such a scenario.

Because time is short, we need a 'no major trade-offs' rebuilding of the economy, and we need to quickly develop the skills and know-how to implement similar, broad-scale decision-making about climate change. Continuing to use negotiation and management methods that routinely result in major compromise and failure will not help.

Past government inaction has also habituated an acceptance of lowered expectations, which has continued to hinder serious climate action. An Australian non-government organisation (NGO) staff member, reflecting on her experiences, said that it has become increasingly clear to her how constrained the environmental organisations are: 'It's a legacy ... they've all come to expect so little environmental responsibility from government, so they don't ask for much in the hope of a small gain. [It's] a very unfortunate situation.'

Timidity, constraint, and incrementalism have, generally, characterised recent federal and state government approaches to environment issues in Australia, and the consequence is that low expectations have become embedded in the relationship between lobbyists and government. When opportunity knocks, or changing evidence demands urgent and new responses, imaginative and bold leadership does not always emerge with solutions that fully face up to the challenge. When, in late 2007, evidence emerged of accelerated climate change, it appeared to have little impact on the climate targets advocated by most of the peak green organisations, which said that their position was 'locked in' till after the election.

Ken Ward, an environmental and communications strategist and former deputy executive director of Greenpeace in the United States, believes that the people who lead environmental foundations and organisations can play a critical part in reconstructing the issue as a climate and sustainability *emergency*—one that takes us beyond the politics of failure-inducing compromise.

With the rapid loss of the Arctic summer ice cover, a climate catastrophe is now in full swing; but Ward says that the opportunity for these leaders to adjust their position is narrow, and this is due, in some part, to the deliberate decision, a decade ago, by environment organisations to downplay climate-change risk. He says:

> [They did so] in the interests of presenting a sober, optimistic image to potential donors, maintaining access to decision-makers, and operating within the constraints of private foundations, which has blown back on us. By emphasizing specific solutions and avoiding definitions that might appear alarmist, we inadvertently fed a dumbed-down, *Readers Digest* version of climate change to our staff and environmentalist core. Now, as we scramble to keep up with climate scientists, we discover that we have paid a hefty price.

For those who have, in the past, downplayed the risks, changing position is now a matter of urgency, because what now needs doing cannot be done incrementally. The desperate measures required to advance a functional climate-change solution at this late date, says Ward, 'can only be conceived and advanced by individuals who accept climate change realities and [who] take the less than 10-year time-frame seriously'. He believes that we need to confront the terror of the situation before we can come to a real solution. 'We are not acting like people and organisations who genuinely believe that the world is at risk. Therefore, we cannot take the measures required, nor can we be effective leaders.'

# CHAPTER 18

# The Gap Between Knowing and Not-Knowing

Why are many powerful organisations, and people in positions of power, so un-terrified, so unwilling to recognise, or advocate, the extent of action now required? The closer decision-makers get to the apex of political and bureaucratic power, the greater the public denial of the need to act at full scale and at great speed.

Much of this incapacity is simply a result of the way power is exercised in societies in which a corporate agenda is the default mode. In the most extreme cases, governments have glued themselves to the fossil-fuel lobby and done almost nothing, even as other significant sectors of business have demanded more action on climate. In his 2007 book, *High and Dry*, former government staffer Guy Pearce documented the influence of the Australian coal industry and the way that big polluters and their lobbyists wrote the Howard government's climate policies to ensure that there was no real plan to reduce emissions.

In politics, 'plausible deniability' has become an art form—a process which ensures that there is no evidence to be found that a person knew, or could have been expected to know, something that may come back to haunt them. A corollary is the pressure for advisors, consultants, and administrative departments to tell a minister or senior official only what they want to hear, or to tell them only the minimum necessary to make a pragmatic decision in the short term.

How many environment or climate ministers in national governments would be able to give an informed and reasonable answer if they were asked to explain the policy relevance of the impending rapid loss of the Arctic sea-ice and its effects on Greenland and sea-level rises?

It is as though people half-recognise the problem, and then deny it if the necessary solutions are beyond their professional boundaries or their perception of 'reasonableness'.

Political leaders now accept that climate change is a problem, and now eagerly embrace small-scale schemes such as changing light globes. Taking the drastic action necessary to turn climate change around, though, is too far outside their political ambit to even consider. Policy analyst George Monbiot noted this limitation: 'When you warn people about the dangers of climate change, they call you a saint. When you explain what needs to be done to stop it, they call you a communist ... everyone is watching and waiting for everyone else to move.'

Psychological denial is the process of refusing to acknowledge the existence, or severity, of unpleasant events or thoughts and feelings: the person keeps on acting as if the event has not occurred. Generally, it is associated with an event or thought whose memory is stressful, so that denial, as a defence mechanism, seems like a good practical strategy; the more dramatic the event, the stronger the denial.

Stanley Cohen, a sociologist at the London School of Economics, has pointed to the paradox of denial: to deny something, you first have to recognise its existence to some degree, so denial is a state of 'knowing and not-knowing'. It can take a number of forms, including outright denial; seeking scapegoats (blaming China, for example); shutting out or suppressing information (so the story is only half-recognised); denying responsibility ('Nothing we can do in this country will make any real difference.'); denying personal power ('No one else did anything, so I didn't either.'); and projecting anxiety (displacing fears onto other issues).

Max Bazerman of Harvard University has asked why societies fail to implement wise strategies to prevent 'predictable surprises'—a term he coins to describe events that catch organisations and nations off-guard, despite necessary information being available to anticipate the event. Think of 9/11, or the failure of American strategy in Iraq. Or climate change.

Bazerman identifies five psychological patterns that help to explain the failure to act on climate:

[P]ositive illusions lead us to conclude that a problem doesn't exist or is not severe enough to merit action ... we interpret events in an egocentric, or self-serving, manner ... we overly discount the future, despite our contentions that we want to leave the world in good condition for future generations ... we try desperately to maintain the status quo and refuse to accept any harm, even when the harm would bring about a greater good [and] we don't want to invest in preventing a problem that we have not personally experienced or witnessed through vivid data.

Bazerman suggests that many political leaders will not want to act until great, demonstrable harm has already occurred.

He also identifies organisational and political explanations for the failure to act on climate: organisational divisions that fail to integrate data, responsibility, and responses; the corrupting power of powerful lobbyists and political donations; and deliberate campaigns to confuse people about the evidence—a tactic long-employed by the tobacco and fossil-fuel industries.

George Marshall, founder of climatedenial.org, says that denial cannot simply be countered with information, because denial is, as Cohen described it, a normal state of affairs for people in an information-saturated society. He also says that the lack of visible public response is part of a self-justifying loop that creates a passive-bystander effect:

[People often won't] spontaneously take action themselves unless they receive social support and the validation of others. Governments in turn will continue to procrastinate until sufficient numbers of people demand a response. To avert further climate change will require a degree of social consensus and collective determination normally only seen in war time, and that will require mobilisation across all classes and sectors of society.

A paradigm shift is also required. We settle into ways of thinking and working that get the job done 'well enough'. As new ideas and changing circumstances challenge our mental model of the world, we may either adapt and change, or stick with what we know and slowly drift away from understanding the world. Even when global warming challenges conventional ways of working, a group of people working together with similar, unrevised mental models will often be able to maintain a degree of social competence and operate effectively with each other. This can be the case even though their collective behaviour no longer fits so well with the conditions of the physical world. Their other, more adaptive, option is to critically re-understand a changing environment, and start to shape and share a new social worldview.

The eminent physicist Max Planck famously observed that 'a new scientific truth does not triumph by convincing its opponents and making them see the light, but rather because its opponents eventually die, and a new generation grows up that is familiar with it'. With climate change, however, we don't have the option of waiting this long.

The complexity and seriousness of climate and sustainability problems makes our current political world of trade-offs, compromises, and decision-making obsolete, along with most of our experience about how to act effectively. This is an extraordinary challenge, because our accumulated skills in the art of compromise become less useful. Perhaps the best way through is to adopt, whatever one's age, a youthful willingness to live with uncertainty and to view the prevention of climate catastrophe as an invigorating process of innovation, learning, and imagination.

Some of the ways that we currently make decisions and plan for the future will also pose big challenges, when our aim is a safe-climate future. An example is how we think about safety itself. Our society has built administrative structures for safety at the micro level — for example, in building design and construction, and in the design and manufacture of aircraft and other vehicles — but no such structures exist for ensuring the safe 'design and development' of the economy, or of our use of resources as a whole. Rather than spending money to test actions in advance, we have considered it more efficient to wait until there are clear signs of a real problem before we take preventive or remedial action. The consequences of this approach are manifest in today's sustainability and climate crises: we fish till the fish run out, irrigate till the rivers run dry, and burn coal till the weather turns hot.

Paradoxically, the more that a society is committed to innovation and to increasing economic output, the more it needs effective processes to ensure sustainability; otherwise, the economy just becomes a machine for testing what we can endure, and we discover what we can't endure only when it is too late to prevent disaster.

Our discussion has been confined to proposing some of the structural and social reasons that people and organisations simultaneously 'know' about the global-warming challenge, but 'don't know' about the speed and depth of action required. This incapacity may be manifested in what we have termed 'blocking': in beliefs that stand in the way of developing necessary solutions. Here are six sets of 'blocking' responses that we have encountered while engaging with people about the sustainability emergency.

The first set that we have often experienced in response to our emergency proposition includes claims that there is no point in taking action, because the climate situation is already a lost cause. Claims include: it's too late, the changes can't be made fast enough, people are too selfish and won't put the needs of others ahead of their own needs, or vested interest will block the necessary action. Other variations include the view that it's hopeless, because our efforts are a drop in the bucket or are insignificant; that people won't change their lifestyles; that governments won't do what is needed; and that people won't accept the government telling them what to do. Finally, there is the view that to succeed we must have action from China, India, Russia, and the US—and since they won't change, we shouldn't either.

These are absolutist arguments: they assume knowledge that nobody can have until after the event. We cannot know whether a mode of action will fail until it is tested, but these propositions, all based on the theme that trying is a lost cause, cut off all possible avenues of success on the basis of vague assumptions rather than detailed knowledge. In fact, recent international examples give the lie to this approach. We reacted quickly and we weren't selfish when the tsunami hit Asia in 2006, or when Burma and China were devastated in May 2008. Why is dealing with climate change more of a 'lost cause'?

The fact is that even vested interests have to operate on the same planet as the rest of us, and vested interests still have children, partners, lovers, friends, and grandchildren. All countries and all people will have reason to act, because climate change will damage them, as well as everyone else. India and China, for example, will suffer from rising sea levels; and the loss of the Himalayan glaciers will be highly disruptive to water supply, through the large rivers that support much of the agriculture in both countries. There is evidence, in other words, to dismantle the credibility of each of these assertions.

A second set of 'not-knowing' responses has been heard many times from leaders of organisations. These include their argument that they can't advocate a comprehensive safe-climate policy because they would lose credibility if they were to 'go out on a limb', or they would not be 'taken seriously' and would, thereby, become ineffective; or that they are 'too busy' to take action on the issue. One Australian non-government environmental organisation recently pioneered a useful way out of this dead end by adopting a two-strand approach. In its public advocacy, it aims for goals and targets that would result in a safe climate as fast as possible. When dealing with governments, though, or other bodies which have demonstrated that they can't yet relate to the safe-climate approach, the organisation will push these institutions as far as they will go. Both strands are acknowledged publicly, and explained. The ultimate objective, with regard to government, is to help them get to a safe-climate goal, too.

In any case, the cost of being forthright is falling. The world's accelerating slide into demonstrably dangerous climate change is now creating new, more favourable dynamics for advocacy and leadership. These days, if one takes a stand that is well based on climate science but which is currently seen as 'extreme', it will be only months, or a year or two at most, before consequences from the real world will show it to be reasonable and necessary. So we can expect that the uncomfortable feeling of being too far ahead of the pack will pass, before too long.

A third form of blocking centres around the proposition that, until we have complete certainty, we shouldn't act preemptively. But when time is running short, and the stakes are high, we simply have to take a risk-management approach and decide to act, despite the uncertainty.

A fourth set of blocking responses is to claim that we can't follow the emergency strategy until other people, or organisations, change their position. But other people and organisations *are* already changing their positions — as witnessed by the increasingly alarmed comments of public figures around the world.

A fifth set of responses revolves around the proposition that a solution to global warming is possible, but only after an unacceptable prior step — in other words, that without change in other significant parts of the system, the level of action required is impossible. Yet there are many examples, including countries going into war, or responding to large natural disasters, which demonstrate that most political systems can switch to an emergency mode when they perceive an urgent need to do so. Variations on this theme include sentiments such as, 'We'd need a huge disaster to happen before people would act', and 'We can't tell the whole truth: we mustn't disempower people.'

Rescuing the planet and getting back to a safe climate will require a huge deployment of physical and economic resources; so, for the duration, keeping jobs and the economy going should not be a problem. And we do not have to wait for a catastrophe — people in many parts of the world are already facing climate disaster. The issue is not whether people have the capacity to engage in powerful acts of the imagination, or foresight, but whether they will apply this capacity in the short amount of time left. Waiting for a climate disaster big enough to motivate action beyond the conventional mode is very dangerous, because by then it may be too late to prevent catastrophic events.

A sixth type of obstacle is the view that a full solution is not possible, so we just have to live with partial failure and partial success. To this end, we've effectively been told: 'The perfect is the enemy of the good—your ideal 'safe-climate' strategy will not only fail but, if it is pursued as the main strategy, it will also block a less desirable, but feasible, outcome.' We are not aware of any credible evidence that a full solution is not possible, or that an attempt to achieve a full solution will make a partial solution impossible. The evidence suggests that the Earth is very close to a period of run-on warming, at which point any partial solution will, fairly rapidly, deteriorate into a much worse state; so, going for the apparently harder safe-climate target now, rather than later, looks like the only practical option.

All these forms of blocking lead to society accepting inaction, or insufficient action, as a solution to the problem. The alternative—to imagine, and plan for, a great transformation of our society in a way that is consistent with a safe-climate future—may be unsettling and challenging. It will require us to change the way we live and how we understand the relationship between our actions and our future on this fragile planet. But it is the only practical option.

# CHAPTER 19

# Making Effective Decisions

In seeking to overcome the advocacy dilemma—the gap between what is reasonable to propose and what needs to be done—four approaches stand out that are relevant not just to policy-makers and lobbyists, but to all of us when thinking about, talking about, and taking action to have the climate crisis recognised as an emergency.

## Pursuing double practicality

The actions that are proposed must be capable of being implemented and, when fully implemented, of *fully* solving the problem. For example, a 3-degree cap may be achievable logistically, but it will not solve the problem of climate safety; rather, it is very likely to result in global warming that is catastrophic and that would escalate beyond our capacity to control or reverse it.

If the world's political leaders work toward a goal that cannot deliver a safe climate, we face a serious environmental disaster. Investments made in good faith in new industrial, urban, and rural systems would also have to be abandoned partway through their economic life, to be replaced by another set of investments made at a time when the environment will be breaking down and the economy will, consequently, be in decline. At such a late stage, the second round of investment could well fail to rescue the situation. The 3-degree cap clearly fails the test of 'double practicality'.

Let's take another example: the idea that we should retrofit coal-fired power stations to run on gasefied goal as a long-term strategy to lower emissions. If a necessary goal for a safe climate is also the complete decarbonisation of electricity generation, this does not fully solve the problem, because it involves building new, large-scale infrastructure that will continue to emit carbon for its investment life. Similarly, proposing to replace gas hot water with gas-boosted solar hot water is not a full solution: greenhouse-gas emissions will still occur when the sun's energy is not enough to keep the water hot and the gas kicks in (on cloudy days, for example). A doubly practical alternative when replacing conventional hot-water systems would be to install a solar unit boosted by electricity from renewable sources.

At a broader level we recognise that, with today's high levels of technical innovation and economic growth, any problem that remains partly unsolved will simply be amplified as the economy grows. So the more committed a society is to economic development, the more it must be committed to *fully* solving environmental problems in an anticipatory way.

### Facing brutal facts

Problems cannot be solved completely if they are not understood completely. When a problem has trivial consequences, a misunderstanding will not matter. But a problem like climate change, which has life-and-death implications, must be understood fully, unobscured by wishful thinking. Winston Churchill, the British prime minister during World War II, understood this. He knew that his strong and charismatic style of leadership might inhibit people from telling him the truth about the progress of the war, so he set up an independent agency, the Statistical Office, to feed him the brutal truth in a constantly updated and completely unfiltered form. Equipped with an unshakeable goal, and a stark understanding of exactly how grim the situation was from moment to moment, Churchill rejected all wishful thinking: 'I … had no need for cheering dreams,' he wrote.'Facts are better than dreams.'

The purpose of facing the facts is not to wallow in anguish, but to inform the creative process so that we can come up with solutions that have the maximum chance of solving the problems, no matter how bad they are. The worse a problem is, the more vitally important it is to know its real nature.

Moreover, we should face all the facts even if we don't, at the start, have all the answers. We do not need a final plan to justify urgent climate action, any more than we would expect a rescue team to have a finalised plan, or a guarantee of success, before they considered attempting a rescue.

Management researcher Jim Collins found companies that survived enormous challenges and continued to thrive were, without exception, the ones that not only had the knack for facing the reality of their situation, but were also driven by an enormously strong will to survive. This led them to develop creative solutions that were equal to the problems.

Collins drew lessons from the experience of Admiral Jim Stockdale, the highest-ranking US military officer in the

'Hanoi Hilton' prisoner-of-war camp during the height of the Vietnam War. Observing his own strategies for survival, and those of his fellow prisoners, Stockdale concluded that 'you must never confuse faith that you will prevail in the end—which you can never afford to lose—with the discipline to confront the most brutal facts of your current reality, whatever they might be'.

This approach combines strategic optimism with tactical pessimism. There is a dogged determination to work for a positive outcome, coupled with an assumption that any number of things can go wrong, unless they are actively prevented.

Collins asks:

> How do you motivate people with brutal facts? Doesn't motivation flow chiefly from a compelling vision? The answer, surprisingly, is, 'No.' Not because vision is unimportant, but because expending energy trying to motivate people is largely a waste of time ... If you have the right people on the bus, they will be self-motivated. The real question then becomes: How do you manage in such a way as not to de-motivate people? And one of the single most de-motivating actions you can take is to hold out false hopes, soon to be swept away by events ... Yes, leadership is about vision. But leadership is equally about creating a climate where the truth is heard and the brutal facts confronted.

Collins' finding is central to climate-policy advocacy: *one of the single most de-motivating actions you can take is to hold out false hopes, soon to be swept away by events.*

To leave people who are facing a serious situation uninformed—when there is a possibility that preventive, or even adaptive, action could be taken—is to leave them living in a fool's paradise. If we do not allow people in on the secret that climate change, if left uncorrected, is going to have disastrous impacts, we get caught in a democratic trap. If political leaders keep problems quiet, they cannot put forward effective policies to solve them. It is doubly impossible for the bigger political parties to provide effective leadership if many scientists, the main sources of objective information, do not emphasise the full range of possibilities, and if environmental and climate lobbyists propose policies that cannot solve the climate problem.

Fortunately, the digital age has helped counter these tendencies by enabling citizens to educate themselves about climate change with a wide array of online resources including dedicated reporting by newspapers and magazines, scholarly papers, videos, and blogs.

## Failure is not an option

When Apollo 13 mission-control flight director Gene Krantz insisted to his ground staff that 'failure is not an option', the meaning was unambiguous: get the imperilled crew home alive, solve every problem faced by the astronauts, and never give up on any life-threatening issue, even when they encountered a dead end. The more difficult a problem, the more effort and creativity needed to be applied, with every conceivable possibility explored until a breakthrough was achieved.

In the case of global warming, the mantra of 'failure is not an option' relates to a broader goal—achieving a safe climate to protect all people, all species, and all generations. Still, the approach is the same. No matter how difficult the problems, society must find a way back to a safe climate. The UK Government's 2006 Stern Review established a powerful justification for this approach when it found that the threat from climate change would become so large that no matter how much it might take to solve the problem, the cost of not controlling it was going to be worse.

Ultimately, this is a moral issue, because the fate of most people, and most plant and animal species, hangs in the balance—determined by whether we can devise and implement a path to a safe climate. We have to find a way to enable the economy to be physically transformed, in the shortest time possible, to safely bring our greenhouse-gas emissions down to zero, to strip excess carbon dioxide from the atmosphere, and to take measures to cool the Earth directly, until a sufficient natural cooling is established. There is no other option that holds out realistic hope of achieving a safe climate.

## Putting the science first

In the case of climate change, facing the facts with brutal honesty requires us to put the science first—which means fully facing up to the ecological impacts of our actions.

Currently, climate policy has been framed as the task of solving the impasse between 'the science' and what is 'politically possible' (which means: what is economically acceptable to governments around the world, who are largely captured by corporate and bureaucratic interests).

The mantra of the former Australian prime minister John Howard was that he would do nothing on climate change that would 'harm the economy', and that it was 'crazy and irresponsible … to commit to a target when you don't know the [economic] impact'. He seemed not to understand that a failure to act would cook the planet. Asked about the impact of rising temperatures, Howard told an ABC interviewer that an increase of 4–6 degrees would be 'less comfortable for some than it is now'. This was a remarkable way of talking about catastrophic climate change. We have a clear responsibility to make politicians put the science first, in the process of framing goals to achieve a safe climate.

Compounding the reticence to take action now is our faith that technology will be able to solve all our problems, including global warming. Post-enlightenment delight at the progress and capacity of technology has produced a cultural impediment to climate action: a technological overoptimism, or determinism, that clouds our understanding that human behaviour needs to change, too.

## Thinking beyond 'business as usual'

One of the things that makes 'business as usual' such a powerful mode of operation is the widely shared assumption that things will go on as they always have. However, given that history has seen many crises — including past climate change, wars and conflict, and the need to build and rebuild economies and societies — it is obvious that things don't always stay the same and that, indeed, we are capable of breaking out of our usual practices when we need to.

When our society has responded effectively to great threats and crises in the past, it has put aside the partial measures and limited possibilities of 'business as usual' and confronted enormous challenges with brutal honesty, finding feasible solutions and pursuing them with single-minded determination. Our preparedness to do so again, when we are confronted with the greatest threat in human history, will determine our success or failure.

# CHAPTER 20

# The 'New Business-As-Usual'

There's a new colour in fashion: warm-climate green. Pastel in tone and hard to miss, you'll find it in newspapers, on television and, especially, in lifestyle magazines, from fashion and travel, to house and garden.

From corporate responsibility to bottled water, climate-friendly images and products reassure us that it is okay to consume as never before. They invite us to feel good at 'carbon neutral' entertainment spectaculars, and to love the celebrities who offset their private jet travel. They invite us to build, drive, buy, fly, shop, eat, drink, and wear sustainability. They assure us that we need new, climate-friendly green things to replace the not-so-green things we have already. But their message is a double-edged fraud: consume even more, and save the planet.

Some of the green-marketing claims are true, within narrow boundaries, but many are not, and only a few paint the big picture of a sustainable path to a climate-safe future. The message is that we can proceed without inconvenience; this is the lifestyle face of the 'new business-as-usual'—an attempt to deal with the immediate pressures of the sustainability crisis in a way that minimises the changes in business models and power relations, at the expense of really solving the problems.

There has been a host of products, services, and market mechanisms developed in response to global warming, but they are not all necessarily about helping create a safe climate. These include 'clean' coal, current-generation biofuels, voluntary carbon offsets, and two arrangements under the Kyoto Protocol: carbon trading, and the Clean Development Mechanism.

## 'Clean' coal

Carbon capture and storage (CCS) is a technique used to remove carbon dioxide from industrial pollution—especially from power stations—and to compress, transport, and permanently store it in secure underground structures, such as expired gas and oil fields, and other geological formations. Spending government money on CCS development is the 'new business-as-usual' mainstay of coal miners, power generators, and the politicians who defend them. But CCS holds out a false promise. At the scale required, CCS is experimental, unproven technology. Further, if it did work, the majority of CCS deployment would not occur until the second half of this century, according to the 2005 *IPCC Special Report on Carbon Dioxide Capture and Storage*. The Australian Labor government's CCS initiative, announced on 25 February 2007, when it was in opposition, envisages the technology only 'entering the grid' by 2030, a timeline that takes it off the table as a near-term emissions-reduction option. It will simply be too late: urgent emission cuts are essential now. If nations in the Asia–Pacific were to adopt a climate-change strategy based on CCS technology, by 2050 emissions would still rise by more than 70 per cent.

While an extensive 2007 study from the Massachusetts Institute of Technology expresses confidence that large-scale CCS projects can be operated safely, it worries that 'no carbon dioxide storage project that is currently operating has the necessary modeling, monitoring, and verification capability to resolve outstanding technical issues, at scale' — in other words, it is not possible to know at this stage if the whole technology-package works. Proposed new plants in Canada and the USA have been scrapped before construction started, largely because they were not cost effective. As a new and complex technology, CCS, like nuclear energy before it, seems destined to be dogged by cost overruns, unforeseen problems, and delays. The biggest concern is that emissions stored underground could slowly leak over time, deferring today's problem to create a monster greenhouse headache in the future.

CCS is inconsistent with a zero-emissions goal because the technology is likely to capture only a portion of greenhouse pollutants, and is energy intensive. It would be possible to capture 80–90 per cent of the carbon dioxide from a coal-fired power station, but only if newly constructed stations were to burn 11–40 per cent more coal to produce the same output. The energy cost would be higher for retrofitted power stations, which have lower CCS efficiencies.

The IPCC finds that CCS would double the cost of electricity where storage sites are distant from power stations.

This would increase the cost of coal-fired power with CCS to more than that of many renewable-energy sources, especially as technology improvements and increasing economies of scale are predicted to halve the cost of renewable electricity generation over the next two decades. Capture expert Greg Duffy told a 2006 Australian parliamentary inquiry that CCS would double the cost of base-load electricity generation, and reduce the output from a power station by about 30 per cent. Lincoln Paterson of the CSIRO told the same inquiry that beyond 100 kilometres, the transport costs may become 'prohibitively expensive'.

A year earlier, a report from five CSIRO energy technology researchers predicted that in five to seven years the cost of electricity from concentrated solar-thermal plants would be competitive with coal-fired generation (without CCS). True to the 'new business as usual' approach, the report was suppressed by the federal government, while hundreds of millions of dollars were allocated for 'clean coal' research. As a result, solar-thermal expertise was driven overseas.

The term 'clean coal' — or, we should say, 'less dirty coal' — also refers to new coal-fired power stations that use Integrated Gasification Combined Cycle (IGCC) technology, a process that first produces a gas from the coal. These plants still emit large amounts of carbon dioxide that would need to be sequestered, their building costs are up to three times that of the most efficient gas-fired installations, and they are more expensive to run than conventional coal plants.

## Current-generation biofuels

Biofuels (ethanol, methanol, and biodiesel) are manufactured from biomass (plant or other biological material) such as crops, or crop and forestry waste, and are considered, by some, to be a sustainable fuel source, because their emissions are part of the carbon cycle. Plants and trees draw down atmospheric carbon through photosynthesis, and the biomass is converted to biofuels, which emit carbon into the air when they combust. This carbon is drawn down again in the next fuel-production cycle.

But current biofuels are manufactured largely from food crops, including maize and soy beans, and from palm oil plantations that are grown in place of rainforests, and this creates its own set of problems. Resolving a multi-faceted sustainability crisis requires an assessment of the life-cycle impact of each proposed solution. This is a test that current-generation biofuels fail.

When the biofuel is derived from broad-acre crops that require nitrogen-emitting fertilisers, such as maize and rapeseed oil, the total energy input can be greater than the output, and the carbon emissions are up to 70 per cent higher than if a car used petrol. In some cases, the biofuel is also not an equal replacement: ethanol burns less efficiently than petrol, for example. The end result is that switching from biofuels back to petrol would produce less global warming; nonetheless, petrol is heavily taxed, while biofuels in many countries are subsidised or taxed at lower rates.

Using crops for biofuels often means converting food sources into energy sources. This transition has seen world food prices double in the five years to 2007, so that under fixed-budget UN food relief programs only half as many people will be fed. World wheat prices, for example, doubled in 2007, and the UN's global food index jumped by more than 40 per cent in a year. US corn farmers, encouraged by government subsidies and rising prices, have turned their fields to ethanol production while, across the border, hunger drove people in Mexico City to riot. As much as 20 per cent of the US grain crop has been diverted to biofuel production, but the quantity of biofuel produced is a substitute for only 2 per cent of the USA's petrol demand.

What's more, if sufficient land were allocated to biofuels to replace current global petrol consumption, there would be no land left for food. If plans to turn more arable land to biofuels collide with a growing population and demand for food, the result will be starvation on a global scale. Swaziland was a case in point: in 2007, while 40 per cent of its people faced acute food shortages, the Swaziland government exported biofuels made from the staple crop cassava.

Using uncultivated land for biofuels also destroys habitats. John Beddington, Britain's chief scientific advisor, says cutting down rainforest to produce biofuel crops such as palm oil is 'profoundly stupid'. In Indonesia, more than a billion tonnes of carbon is pouring into the air each year as thick rainforest is cleared for cropping; still, the country plans to expand palm oil production to 260,000 square kilometres by 2025.

'The competition for grain between the world's 800 million motorists, who want to maintain their mobility, and its two billion poorest people, who are simply trying to survive, is emerging as an epic issue,' says Lester Brown, of the Washington-based Worldwatch Institute, who notes that, in seven of the past eight years, the world has grown less grain than it has used, so that the world's grain-stock reserve was down to 50 days by the end of 2007.

In 2007, the UN special rapporteur on the right to food, Jean Ziegler, denounced biofuels as 'a crime against humanity' and called for a five-year moratorium on their production. If that doesn't occur, says policy analyst George Monbiot, 'the superior purchasing power of drivers in the rich world means that they will snatch food from people's mouths. Run your car on virgin biofuel and other people will starve'.

The quantity of biofuel that can be produced in a sustainable manner is also likely to be very small compared to current demand for petrol, and its current production is a very narrow response to peak oil.

Can biofuel production ever be sustainable? It depends on the source of the biomass, and how and where it is produced. Second-generation biofuels made from wood, straw, or waste from agricultural cropping will become commercially available. They have the capacity to complement sustainable agriculture and forestry practices, and to be co-produced with agricultural charcoal to sequester carbon. To that extent, biofuels have a future; but not when rainforests are destroyed, biodiversity is decreased, food production is lost, and small landholders in the developing world are forcefully displaced.

## Voluntary carbon offsets

Carbon offsetting means that emissions from household utilities, transport, or commercial activity are balanced by buying a product that will reduce emissions elsewhere, or reduce greenhouse-gas levels. The carbon-offset product may be an investment in a program that will draw down carbon, such as tree planting, or a project that will reduce future carbon emissions, such as achieving energy efficiency or building renewable-energy capacity. But when all sectors of the economy require deep and urgent emission cuts, as our current climate emergency demands, we all have to play our part, rather than paying someone else to do it.

As a commercial product, carbon offsetting has a potentially dangerous effect on people. A good analogy of this effect is the medieval Church practice of selling indulgences to sinners in order to lure them to buy absolution. As sinners bought absolution, so they were free to sin again—just as buying offsets assuages people's guilt about producing carbon emissions. Too often, offsetting is an eco-fantasy that justifies a high-carbon personal or corporate lifestyle.

In the competitive commercial world, carbon offsetting also risks becoming just a cheap publicity stunt to push the appeal of a new album or concert tour. In this form, carbon offsetting encourages complacency, displaces real actions, and fosters the illusion that we can keep on polluting forever.

A *Financial Times* investigation published in April 2007 found that companies and individuals rushing to go green 'have been spending millions on carbon credit projects that yield few if any environmental benefits'. It uncovered widespread failure in the new carbon-offset markets, suggesting that some organisations are paying for emissions reductions that do not take place, while others are making big profits for very small expenditure and, in some cases, for clean-ups that they would have made anyway. It also found carbon-offset selling services of questionable or no value, and a shortage of verification.

Distorted economic relations across the world can also undermine the value of offsets; for example, offsets may be exported to developing countries, where costs are lower and the balance is pocketed by the carbon-offset entrepreneur. In other cases, offset schemes are just not viable: in one example, a British company bought treadle water-pumps to replace diesel pumps for Indian farmers, in order to reduce local emissions and thus 'offset' Westerners' air travel. The reality on the ground is that if a peasant farmer treads for two hours a day, it would take at least three years to offset the carbon dioxide from one return flight from London to India—luxury travel is 'offset' by Indian human energy.

The best offset schemes give a guaranteed result by investing in renewable energy to reduce emissions, and are effective in acting as a social-change agent by building infrastructure and by encouraging policies that will cut future emissions and are consistent with the need for a zero-emissions economy. These are not the cheapest schemes.

Other schemes may be genuine, but misguided. Trees take years to sequester carbon after they are planted, so reafforestation offsets are doing very little to reduce global warming now, when it really counts, and are difficult to verify. Trees don't offset anything if they die from changing rainfall patterns or neglect, and in many cases the effect of tree planting is only to pay back the carbon debt incurred when the land was first cleared.

And at the cheap end of the offset market are cowboy operators whose schemes lack transparency: trees may not be planted, or may be counted multiple times, or may be paid for by government grants and then resold as offsets. In an industry where there are no widely accepted standards or verification procedures, there is no accountability for this sort of activity.

Some travel-offset schemes promoted by airlines greatly underestimate the impact of the flight, because they fail to account for the fact that emissions at high altitude have almost three times the effect than they do at ground level.

On the other hand, some people who have made a real effort to reduce their emissions as far as practicable have found that there is still an emissions gap that they want to address, and they have found well-designed schemes that will structurally reduce emissions production.

In the end, for carbon offsetting to work, its market needs to be strongly regulated to ensure honesty, accountability, and verification, with appropriate technologies and schemes that encourage behaviour that is consistent with achieving a safe climate.

### Clean Development Mechanism

Currently, the biggest carbon-offset scheme is the Clean Development Mechanism (CDM), which was established by the Kyoto Protocol and has been in operation since 2001. Under the scheme, wealthy nations that are required to cut emissions under Kyoto can get credit by investing in large-scale projects in the developing world, where it is generally cheaper to achieve the same amount of emissions reduction. Organisations can buy Certified Emissions Reductions (carbon credits from these projects) to meet their national or regional offset carbon-reduction obligations. In theory, emissions-reducing projects in developing nations must be verified as being genuinely new activities that would not otherwise happen without the funding.

This scheme was exploited from the start. In March 2007, *Newsweek* reported: 'So far, the real winners in emissions trading have been polluting factory owners who can sell menial cuts for massive profits and the brokers who pocket fees each time a company buys or sells the right to pollute.' An investigation by Nick Davies of the *Guardian* found that the

CDM had been 'contaminated by gross incompetence, rule-breaking and possible fraud by companies in the developing world, according to UN paperwork, an unpublished expert report and alarming feedback from projects on the ground'. In one instance, carbon offsets for a US$5 million incinerator in China that was built to burn, rather than emit, hydrofluorocarbon gases were sold to European investors for $500 million.

Half of the offsets certified under the CDM in the initial period were for five similar large projects in India, China, and South Korea, where over-priced credits were sold for many times the cost of the action. In many cases, it was hard to demonstrate that emissions would be reduced, or to verify the amount. There was also evidence that as many as one-fifth of projects had been wrongly checked, and that many projects are blatantly 'non-additional'; that is, they would have gone ahead regardless of the CDM, and do not represent real additional emissions reductions.

Instead of stimulating new investment in the best green technologies, such as renewable energy, the CDM has mainly granted carbon credit to projects that would have been built anyway, such as large hydro and wind projects. A December 2007 study of the 654 hydro projects at various stages of the CDM approval process found blatant and widespread non-additionality. More than one-third of the large hydro-electric schemes that had been approved for credits were already completed before CDM approval; the majority of the projects (89 per cent) were expected to be completed within a year following approval; and almost all (96 per cent) were expected to be completed within two years. If you consider the long lead times for hydro construction, it becomes obvious that these projects were going to happen anyway, and that the many millions of credits that they generate will merely allow industrialised countries to meet their targets without reducing emissions. Further studies have confirmed that projects that use other technologies, such as wind, also suffer from widespread non-additionality.

## Carbon trading

CDM offsets are one element of the larger carbon market that has been set up by the Kyoto Protocol under the United Nations umbrella. Carbon trading is another, which is supposed to be an enforceable mechanism for reducing emissions. Under carbon trading, a total emissions target is set for an industry, or region, and is decreased over time. Quantity permits that are equal to the target are sold, and emitters must buy permits to match their level of pollution. In the name of efficiency, permits are traded. Over time, the number of permits is reduced, and their price increases due to increasing scarcity. As a result, the incentive to switch to low-pollution technology increases.

Can't go wrong? The carbon-trading market for Europe, known as the European Union Emissions Trading Scheme (EUETS), got off to a very bad start. The initial permit pool was too large, because of business lobbying, and permits were given away as rewards to the biggest polluters. These businesses then realised that they had more permits than they needed, so they sold them at huge profits. When everyone realised what had happened, the price of permits collapsed. As the price collapsed, so did the impetus for some viable CDM projects.

As a result of the scheme, some of the biggest polluters

earned hundreds of millions—much coming from the budget of public institutions, including universities and hospitals which had to purchase permits—and emission cuts were displaced onto the developing world.

Most current carbon-trading schemes have deep structural flaws: permits are given away to the biggest emitters, and pollution is transformed into a private property right; the need for deep emission reductions in the highest-polluting rich countries is shifted to developing nations; targets are inadequate, and verification and enforcement often poor; and money and effort is poured into trading carbon and finding loopholes, rather than into renewable energy.

Carbon trading also encourages the lowest-cost choice to the detriment of other factors: electricity generators may decide that switching from coal to gas fits the scheme's criterion, while the social imperative would be to invest in renewable-energy capacity to develop the technology, build productive capacity, and reduce the cost.

Another structural flaw to carbon trading is that current systems don't include shipping and air-travel emissions, which are two of the fastest-growing emission sectors.

It's hard to avoid the impression that many of the passionate advocates of carbon trading see it as a way to make a great deal of money from the process of trading pollution rights, rather than as a means to cutting emissions to zero, so that greenhouse-gas-emitting technologies become obsolete. As a September 2006 report from the Dag Hammarskjöld Foundation concluded, 'With a bit of judicious accounting, a company investing in foreign "carbon-saving" projects can increase fossil emissions both at home and abroad while claiming to make reductions in both locations.'

So does carbon trading have any role to play? It is not difficult to design a system that avoids most of the pitfalls mentioned. It must cover all emissions, have a sharp and clearly defined declining cap that fits the need for a rapid transition to a safe-climate economy, and it must include border protection, to stop responsibilities being exported to low-wage countries. Such a system would be better called 'cap and auction' rather than 'carbon trading', to emphasise its intended purpose. There are also compelling reasons why 'cap and auction' schemes should be kept within national borders, especially for high-polluting developed nations, so that emissions are cut *within* that economy, rather than the buck being passed to a less-developed country.

Carbon trading has a part to play in our climate emergency, but it is not the main game. The emergency requires strong regulation and intervention in the market, which cannot respond by itself at the depth and speed required. It is also necessary to develop coordinated plans to build renewable-energy capacity and to improve energy efficiency, along with allied regulations to step down the use of coal and gas.

In the past, rule-based methods for reducing environmentally damaging substances, such as lead in petrol, were effective; but cutting total carbon emissions is more challenging, because fossil-fuel use is ubiquitous. Since the quantities of petrol and natural gas, for example, are subject to a declining cap as part of an emissions-reduction strategy, how do we allocate the right to their use between competing householders and other users? A tax rate may work, but it would be chronically inequitable. Rationing would be fairer, but a black market would emerge if the trading of rations did not have a legal and regulated basis. Carbon trading will occur, one way or the other, but it is not the primary strategy to rely on as the climate crisis deepens.

While the 'new business-as-usual' mode is, in many cases, a well-intended response to the emerging climate and sustainability crisis, it still involves avoiding the deeper nature of the crisis. The question then becomes: can we be brutally honest and doubly practical, and still get beyond 'business as usual', in either its traditional or new form, to build a truly sustainable society?

# CHAPTER 21

# Climate Solutions

If you look at many of the actions being taken around the world today—such as increasing energy-efficiency, building large-scale renewable-energy plants, or moving from private to public transport—it is easy to see that many perfectly workable solutions to global warming exist now. However, they are not the whole answer.

Climate solutions must complement solutions to other sustainability problems, such as water, food, and peak oil, at the same time as cooling the Earth by at least 0.3 degrees, compared to the present, as fast as is practically possible. To make this possible, three types of action are required.

The first is to stop adding to the upward pressure of temperatures by cutting human greenhouse-gas emissions to as close to zero as possible.

The second is to reduce the amount of carbon dioxide in the atmosphere, because even its present level will push temperatures higher than they are now, and will trigger positive feedbacks that add further warming. This may be achieved by allowing natural carbon sinks to draw down excess carbon dioxide from the air, and by adding large-scale human processes for capturing and storing excess carbon dioxide from the air. As an example, we could grow trees and other biomass on a large scale to make agricultural charcoal, which can be stored in soils while also adding to fertility.

The third action is to consider environmentally safe options that produce negative forcings, or change the energy imbalance in the climate system, to slow the peak rate of warming and bring the global cooling forward to the earliest possible time. Examples include promoting forest growth (which helps stimulate the creation of clouds), and increasing the growth of ocean algae (which draws down carbon dioxide and also helps stimulate cloud production). Other options that are still being studied to determine their environmental and technical suitability include the notion of seeding the stratosphere with sulphates to create a reflective layer.

## Stopping additions to heating: cutting emissions to zero

If you consider the range of sustaining technologies that has been created over the last 30 years through human innovation, it is clear that the options available for drastically cutting greenhouse-gas emissions are not beyond our collective capacity or imagination. *Zero Carbon Britain*, an alternative energy strategy that was released in 2007, finds that in 20 years the UK could produce 100 per cent of electricity without the use of fossil fuels or nuclear power, while also almost tripling electricity supply, and using it to power most heating and transport systems. A similar strategy for the United States is *Carbon-Free and Nuclear-Free: a roadmap for US energy policy*, a joint project of the Nuclear Policy Research Institute and the Institute for Energy and Environmental Research, which was published in 2007.

The failure, so far, to engineer energy use along sustainable paths is not a failure of technological or economic capacity, but of political and social will.

There are new, lightweight materials for vehicle construction, and household appliances that use a small fraction of the energy of those now in use. Carbon-neutral buildings do work, electricity from renewable sources is the fastest-growing energy industry, and hundreds of millions of people are moved by electric mass-transport every day. With high-speed electric rail and advanced telecommunications, we could manage without mass air travel. One study has estimated that the savings that would result from using telecommunications networks to conserve energy and to increase clean-energy use at home, in the workplace, and in the ways we connect people to be 5 per cent of total Australian emissions, with estimated energy and travel-cost savings of A\$6.6 billion per year.

The key strategies to cut greenhouse-gas emissions to zero are resource efficiency, backed up by substituting renewable energy for fossil-fuel sources. The integration of these strategies is illustrated in two sectors: materials production, and transport.

## Efficiency
The greatest reduction in greenhouse emissions — and the most economically efficient — can be made through comprehensive, visionary efficiency programs for energy and other resources.

Investing in resource efficiency — which cuts the amount of materials needed to meet human needs — produces many side benefits, such as less ecological damage, less resource depletion, and fewer adverse impacts on human health. The greater the level of efficiency, the greater the benefit.

In California, energy-efficiency programs that have been implemented over recent decades have held electricity consumption per capita, roughly, at a constant, while overall per capita US consumption has almost doubled. A *McKinsey Quarterly* report says that a 50 per cent cut in energy use is feasible, using off-the-shelf technology. If we knew that the price of energy would double in say, five years, we could almost certainly double our efficiency.

McKinsey & Company's analysis for Australia, 'An Australian Cost Curve for Greenhouse Gas Reduction', was released in February 2008. It found that cutting emissions by 30 per cent by 2020 and 60 per cent by 2050 is achievable without incurring a major impact on consumption patterns or quality of life, and without major technological breakthroughs or lifestyle changes (by 'using existing approaches and by deploying mature or rapidly developing technologies to improve the carbon efficiency of our economy'). It estimates that by 2020 about 25 per cent of the total reduction can be realised with positive returns (actions that save, rather than cost, money): 'Most of these beneficial (or 'negative-cost') opportunities are energy-efficiency measures related to improvements in buildings and appliances.' Overall, the analysis found that the 2020 target could be achieved at an average cost of A$290 per household, compared to an estimated increase (by McKinsey) in annual household income of more than A$20,000 by 2020.

Friedrich Schmidt-Bleek founded the Factor Ten Institute to provide practical support for achieving significant advances in sustainable production, in particular through increases in resource productivity throughout the economy. He says that an overall energy-efficiency improvement of up to 90 per cent is achievable with current commercially available technology.

The scandal is that sometimes the most energy-efficient domestic technologies and appliances are not even available in many countries, or businesses and consumers are not aware of their availability. On average, a refrigerator in the USA uses double the electricity of a refrigerator in Europe, which, in turn, uses four times the electricity of the most efficient refrigerator on the market (made in Turkey, and currently not even available in Australia). Compared to typical refrigeration in use today, leading-edge fridge technology with vacuum panel insulation can reduce energy needs by 80–90 per cent, can cut typical peak loads by 100 watts per unit, and can avoid supply-side investment (in generating capacity) of A$200 per household, or A$1.5 billion for Australia as a whole.

World electricity demand could be cut by 25 per cent just by introducing market-leading appliance and lighting efficiency standards, while zero-emission homes and commercial buildings are now a reality. The UK government has legislated that from 2016 all new homes are to be zero emitters for heating and cooling, while large eco-towns are already being planned. The French government has made a commitment that all new buildings will be net energy producers by 2020, and the German government has a 20-year program to upgrade the nation's housing stock to meet high energy-efficiency standards.

## Renewable energy

A range of renewable-energy technologies is now available for power generation, and University of New South Wales researcher Mark Diesendorf says there is no technical reason to stop renewable energy from supplying all grid electricity. Is such a technological turnaround feasible in a short period of time? Rapid economic changes do happen surprisingly often. Between 1986 and 2001 the annual production of mobile phones rose from one to 995 million. Today, 160 million people in China get hot water from solar water heaters. One-third of the world's installed solar-panel capacity is on German houses, because of far-sighted government policies.

In Europe, wind power is widely utilised, and generating costs are falling. By 2020, wind is expected to be a competitive primary-energy supplier, whether or not there is a price on carbon emissions. In Denmark, the government plans to generate 75 per cent of national electricity needs through wind power by 2025. By 2010, Germany will have installed wind-power potential sufficient to generate the equivalent of 40 per cent of Australia's current electricity needs.

As the scale of production increases and costs continue to decline, solar photovoltaic (PV) energy may become the cheapest source of energy in many locations, because it can bypass ageing and fragile electricity grids and deliver power directly to the end user, fundamentally changing the underlying economics of energy. Germany's PV revolution means that more than 400,000 German homes have installed solar panels; with the current growth-rate of installations, Germany plans to be installing over one million solar electric units per year on house rooftops by 2010. As the scale of PV increases and innovation continues to reduce panel prices by, perhaps, half in the next decade or so, an energy-efficient Australian home might be made, essentially, self-sufficient in electricity, for only A$10,000 to A$15,000. This is comparable with the current cost per house of building generating and power-line electricity supply, using large, centralised, coal-fired power stations.

When the cost of coal-fired electricity increases, because of carbon-emissions pricing and trading, the economics of household PV installations will improve further. The next generation in solar technology, developed in Australia, is the sliver-cell solar cell, which is more efficient while using less silicon, which is very expensive. This would bring the cost down dramatically and revolutionise the uptake of PV solar energy, but its development is currently stymied by a lack of research support from the government.

By far the lowest cost option for solar electricity is solar-thermal technology — or concentrating solar thermal power (CSP) — which uses the sun's radiation to heat fluids that carry the thermal capacity to generators. The best-recognised installations of CSP, to date, are in the Californian desert, where extended rows of cylindrical-parabolic collectors concentrate the sun's rays towards long, fluid-filled pipes.

The Club of Rome, the global think-tank and centre of innovation that produced the agenda-setting report *Limits to Growth* in 1972, has now joined with the Trans-Mediterranean Renewable Energy Cooperation to propose a bold new energy scheme for Europe that uses CSP. The new proposal is for the fast deployment of CSP technology in desert areas of North Africa and the Middle East, and to link the European, North African, and Middle Eastern electricity markets, using new technology and low-loss transmission grids, which will be able to supply 90 per cent of electricity requirements. Such 'additional strong and determined emergency measures' are now required, they argue, because 'it is now too late to achieve the required U-turn with a business-oriented slow transition to low/no carbon technologies'.

The technology is on the cusp of some remarkable scale-up breakthroughs, such that it is predicted to be cheaper than coal within five years. An area of solar thermal collectors that is 35 kilometres square in a high-irradiance area would produce enough electricity to meet Australia's total power needs.

There is also great capacity in Australia for generating bioelectricity (electricity derived from biomass). A September 2007 report found that by 2020 bioelectricity could deliver the equivalent of 8 per cent of the electricity generated in 2004, with most biomass coming from 'wheat stubble, plantation forest waste, sugar plantation waste, and oil mallee'. The report explains that these are promising sources, from which 'no land is transferred from food production to bioenergy production. Indeed, oil mallee can help to combat dry-land salinity and hence will make more land available to food production'.

Other renewable technologies—such as geo-thermal energy, and wave and tidal power—will all play growing roles in eliminating the use of fossil fuels in the economy. Iceland, for example, now heats close to 90 per cent of its homes with geothermal energy. With economies of scale, continuing innovation, the introduction of a reasonable price on carbon emissions, and the impact of climbing oil prices, renewable-energy technologies will become the most cost-effective means of producing electricity.

## Materials production

In industry, efficiency programs are reducing greenhouse-gas emissions from energy use by as much as 80 per cent, by using smart technologies, new processes and materials, co-generation, and by relocating production. Increasingly, lower-impact substitutes are being produced for materials such as aluminium, cement, and steel, which require large energy inputs during production. Alternatives to traditional cement and concrete, for example, include geopolymers (alumino-silicate products created from clay-like materials); magnesium cement; fillers, such as ground-waste glass, instead of concrete; the use of lightweight construction techniques; and additives that increase strength and allow a lower volume of concrete to be used. In both the aluminium and steel industries, options include greater use of recycled materials, carbon-fibre composites, high-strength alloys, and optimising design to reduce the quantity of material required.

Five broad strategies can be applied to the energy-intensive materials sector, especially to metal and cement production.

The first is to redesign the products and the platforms that deliver services, along with their associated supply chains, so that their production and use *needs less material*. Designing products, buildings, and infrastructure to use less material depends on a range of strategies, such as enabling long life, re-use, and effective maintenance and repair, and also 'lightweighting', which is the strategy of using the minimum amount of material and weight necessary to achieve a structure or purpose. Lighter vehicles, for example, use less fuel, lighter buildings need less material to support, and lighter products need fewer resources to manufacture, and less effort to dispose of them.

Lightweighting has been employed increasingly since the 1970s oil crisis. Previously, a heavy weight was thought to correlate with strength, reliability, longevity, and quality; now, a lighter weight is thought to correlate with sophistication and quality.

The second is to *recycle* materials in an energy and materials-efficient way, to recover discarded materials. Nature's systems have evolved to achieve extraordinarily high recycling rates. For example, carbon recycling in natural systems is more than 99 per cent. The scale of the human economy is now so huge that the same imperatives to recycle exist; for example, the explosive growth in the production of mobile phones is leading to a critical shortage of the geological reserves of some of the rare minerals used in the advanced electronics. Recycling is beginning to look like the only way to keep the mobile phone sector viable. Generally, much more energy is required to create virgin resources than to create recycled resources. To tackle the climate issue, a great deal of physical transformation needs to occur—inefficient buildings, cars, products, and infrastructure will need to be retrofitted, or replaced, in a relatively short period of time. This could involve a large new burst of carbon dioxide release, unless old materials are efficiently recovered from the scrapped assets and are recycled into new assets.

The third strategy is to *substitute* materials that have lower-embodied climate impact (where this improves the performance of the system as a whole). Substitution can make big cuts in greenhouse emissions; for example, by substituting geopolymer cement, very high extender-content cement, and magnesium-based cement for traditional calcium-based cement, which is very energy-intensive to produce, accounting for around 4 per cent of global carbon dioxide emissions. As another example, in areas in which lightweighting is critical, like car manufacture, steel can be replaced with carbon fibre from renewable sources. 'Petroleum' products (a range of chemicals and plastics) made from compounds sourced from plants can also be substituted in place of fossil-fuel-derived petrochemicals.

The fourth strategy is to *switch energy sources* used in the production of each particular type of material. Remote-area mining and mineral processing operations are beginning to identify opportunities to use solar, wind, and geothermal energy. Australia faces a strong challenge in the production of aluminium because it is one of the few places in the world where the industry is almost entirely dependent on electricity from coal-fired power stations. Elsewhere it is largely produced using hydro or geothermal power or fossil-fuel gas power. As the world shifts away from fossil-fuelled energy, all materials will be made using climate-safe energy sources.

The fifth strategy is to use long-lasting materials as a way of *sequestering excess carbon* from the air. In some situations it is possible to sequester some of the excess carbon from the air into materials that can be recycled 'endlessly', or into products that have very long lives. Carbon dioxide from the air, for example, can be trapped through plant growth, and the plant material can be processed to make char, which can replace carbon from the coal used in steel manufacture. In this way, steel becomes a storehouse for excess atmospheric carbon.

In practical settings, these five strategies are usually combined to create the maximum impact. During the transition to a safe-climate economy, it will be important to coordinate the changes so that perverse results are not produced. Lots of cement and concrete are needed during an intense structural-change period, for example, and it would be counterproductive if we were to cause greenhouse-gas emissions to rise rapidly while we attempted to replace inefficient or inappropriate products, buildings, and infrastructure according to an accelerated schedule. The potential impact could be cut by ensuring that any increments to energy supply are from renewable-energy sources, and by preparing to switch as early as possible to low-impact sources, or types, of cement, concrete, steel, aggregate, and so on. To make sure that the transition is as effective as possible, it needs to be planned, and modelled to identify opportunities for synergy.

## Transport

Our aim must be to eliminate all fossils fuels from the transport sector. We already have the technology to electrify rail networks, which means we could shift freight from roads, and people from planes, using renewable energy—a zero-emission form of transport. In France, new road construction is being severely curtailed in favour of expanded rail travel that uses state-of-the-art fast-train technology. Already, this is replacing air travel on many routes, because of shorter travel times, reduced check-in security and formalities, and the convenient location of stations in city centres. The high-speed rail link between London and Paris now takes less than 140 minutes. You can end up waiting in an airport for that long when you fly. Five years of Australia's major road funding would be enough to electrify a basic national rail network for freight. A high-speed passenger service would require new infrastructure.

New car technologies are also enabling fossil-fuel reduction. Commercially available hybrid vehicles use about half the petrol of a similarly sized car, while plug-in hybrids use a quarter of the petrol. New lightweight steel substitutes have created further efficiencies, allowing a total reduction of 90 per cent in fuel use. Fully electric vehicles are now also available, and their costs will go down with increasing scales of production and with further innovation. Honda has unveiled a zero-emission fuel-cell vehicle with a top speed of 160 kilometres per hour and a range of about 430 kilometres. Electrically assisted bikes also achieve very impressive environmental performance, while a normal bicycle, which requires 99 per cent less material and construction costs than a car, is an extremely efficient option. Further changes to urban layout to create hubs, and to increase the density of buildings, would also allow cities to use walking as the principal means of mobility, with public transport and bicycles as dominant support-modes.

## Reducing the current heating processes

To reduce atmospheric carbon, the Earth's natural carbon sinks must be protected and strengthened. Processes that isolate and draw down, or sequester, atmospheric carbon into the ground have become of great interest to scientists. Human activity can actively expand these processes. To be effective, they must be substantial in their long-term outcome, and have a low risk of failure, or rapid or large-scale leakage.

There are a variety of measures that can enhance these natural sinks, including protecting and expanding forests.

Recently, attention has been drawn to *Terra preta do indio*, or 'black soil', a term that describes dark, rich Amazonian soils, some as old as 7000 years, which contain many times the amount of carbon found in the soils of the surrounding areas. Once a mystery, it is now believed that the soils were deliberately fertilised and enriched by the region's original human inhabitants, using charred waste that was buried and maintained in the Earth.

Today, 'black soil' (or *terra preta*) refers to land enriched with carbon, and other minerals produced from biological materials, through a process called pyrolysis. Biomass, such as crop residue or wood, is transformed into agricultural charcoal, or biochar. This carbon-rich material is buried in the soil, where it may effectively sequester atmospheric carbon for hundreds, and even thousands, of years.

Biochar has caused great excitement among climate scientists because of its ability to take greenhouse gases out of the atmosphere and reduce greenhouse-gas concentration. Its benefits as a soil additive include mineral and microorganism enrichment, and increased water retention, and it is especially useful in countries such as Australia, where three-quarters of the soils contain less than 1 per cent carbon. Bruno Glaser of the University of Bayreuth in Germany has found that crop productivity can double in *terra preta* soils. The soil's increased fertility, and better ability to tolerate weather extremes, result in higher plant yields and nutritional content, and allow a move from the current dependence on industrial fertilisers to an organic method of enriching and restoring farming lands. A 1-metre-deep hectare of *terra preta* can hold 250 tonnes of carbon, compared to 100 tonnes from unimproved soils. This is more effective as a carbon-sequestration technique than growing forest; and, unlike forest, there is less risk of fires producing large-scale releases of the sequestered carbon.

Joe Herbertson of sustainability consultants Crucible Carbon describes his reaction when he read about biochar technology: 'The hairs went up on the back of my neck ... this is the best news on climate change I've ever heard.'

The process that generates the biochar that is used to make black soils (pyrolysis) is, optimally, an anaerobic (oxygen-free) thermal process in which biomass is baked in a kiln to produce charcoal. Useful byproducts include bio-oils and various fuel gases, such as methane and hydrogen, that can be used for combustion, or to feed fuel cells.

Today, biomass is being used in other ways for power generation, but research and development has demonstrated the feasibility of biochar sequestration as a realistic means of reducing carbon dioxide levels. The big question is whether the *terra preta* process is cost effective in drawing down atmospheric carbon, and whether it can be done on a large-enough scale to have a significant impact.

The practicality and economic viability of the process in reducing carbon dioxide levels to at least 350 parts per million has been explored by Michael Obersteiner of the International Institute for Applied Systems Analysis. He suggests that a maximum removal-rate of five billion tonnes of carbon a year would mean that, assuming no other emissions take place over this period, all historical emissions could be reversed in 70 years.

Johannes Lehmann of Cornell University estimates that *terra preta* schemes, working with biofuel production, could store up to 9.5 billion tonnes of carbon a year—more than is emitted by all of today's fossil-fuel use. Nonetheless, as we've seen, biofuel programs today are having widespread detrimental effects, including the transfer of land from food to fuel production.

The introduction of large-scale *terra preta* would need to be balanced with sustainability needs, so as not to become another tool for the corporate destruction of traditional farming practices, especially in the developing world.

Nobody wants biochar to require such large areas of new monocultures that it would end subsistence farming and the use of land for pasture, which would produce a shift to industrial agriculture worldwide and turn natural forests into vast industrial tree plantations. Still, biochar has a place in developed and developing worlds, provided there are strong land-use controls. While traditional charcoal kilns are inefficient and produce harmful emissions, including black soot, modern small-scale pyrolysis units can be used appropriately with traditional agriculture. If slash-and-burn is replaced by slash-and-char, up to 12 per cent of total human carbon emissions by land-use change can be offset.

There is another benefit, too. Nitrogen fertilisers are a major source of the potent greenhouse gas nitrous oxide, but char-enriched soils have shown a 50–90 per cent drop in nitrous oxide emissions, as well as reduced phosphorous runoff.

As *terra preta* enriches the soil, less land will be needed to produce a given amount of food. Even if we deployed only existing crop waste, we could remove about one billion tonnes of carbon from the atmosphere per year. If degraded waste and unused croplands were added, it may be possible to double that figure. In May 2007, the conference of the International Biochar Initiative was told that, as well as producing biofuel, biochar could produce a 'wedge' of carbon reduction amounting to a minimum of 10 per cent, and possibly much more, of world emissions.

Biochar technology does away with the trade-off that we are witnessing with broadacre biofuel crops, which enforces a choice between fuel and food. The biochar method can use crop waste, degraded lands and, perhaps, a proportion of plantation forest residue (provided that forest regeneration is not impeded) to enrich soils, thereby increasing food production, as well as sequestering carbon. Like energy efficiency, it is a solution that more than pays for itself.

# CHAPTER 22

# Can 'Politics as Usual' Solve the Problem?

It would be unsurprising to discover that many people perceive that the gap between what is being done and what needs to be done about global warming is growing bigger. There is increasing public unease, matched by a political incapacity to publicly recognise the true scale of action that is now desperately required.

Reflecting on his experience as a young man in Britain in the late 1930s, environmental scientist James Lovelock says:

> [I'm] old enough to notice a marked similarity between attitudes over sixty years ago towards the threat of war and those now towards the threat of global heating. Most of us think that something unpleasant may soon happen, but we are as confused as we were in 1938 over what form it will take and what to do about it. Our response so far is just like that before the Second World War, an attempt to appease. The Kyoto agreement was uncannily like that of

Munich, with politicians out to show that they do respond but in reality playing for time.

We have procrastinated for so long on global warming that it is now essential to move to a safe-climate economy as fast as is safe and practicable. A quick sketch of that task would include:

- building the capacity to invent, plan, model, and coordinate the new economy, and encouraging research to develop and scale up the new technologies and products;
- building the physical infrastructure and capacity to produce safe-climate goods and services, such as new production lines for ultra-energy-efficient home appliances, and zero-emission vehicles for public and private transport;
- developing national energy-efficiency programs for building and industry, with enforceable minimum standards. Assistance would be given to householders, especially those on lower incomes, to reduce energy use;
- constructing large-scale renewable-energy plants and local and household energy systems to allow the closure of the fossil-fuel-fired generating industry. This may include very large solar-thermal projects and wind-energy schemes.
- upgrading and electrifying the national and regional train grid so that long-distance road and air freight can be shifted to rail;
- providing safe-climate goods and services, expertise,

and technologies to less-developed nations to support their transition to the post-carbon world;

- providing adjustment and re-skilling programs for workers, communities, and industries affected by the impacts of global warming and by the move to the new economy; and
- developing the bio-char and re-afforestation industries.

Speed is essential — the emphasis is not just on building a safe-climate economy, but on doing so as quickly as is feasible while protecting the environment and keeping the rate of temperature change at a safe level. If the world takes ten or 15 years to stop increasing the rate of emissions, and another 40 or 50 years to stabilise atmospheric carbon levels, it is very likely that the resulting warming (an increase of more than 2 degrees), and its rate of increase, will be too much for many ecosystems, and may trigger positive climate feedbacks that escalate warming beyond control.

Peak oil, in the end, may contribute to lowering emissions. As the cost of petrochemicals is driven higher, the way that we produce and consume goods will change. Rising transport costs will put downward pressure on the global distribution of lower-value goods per weight, and more goods will be produced locally, which will redistribute global manufacturing capacity and jobs. A 25 per cent increase in fuel prices produces a 10 per cent increase in freight rates, reducing international trade by 5 per cent, according to Thomas Homer-Dixon, author of *The Upside of Down*. It is reasonable, then, to expect that if the price of oil doubles it would cut international trade in material goods by one-fifth. We can expect that 'food miles' and 'product miles' will influence consumption patterns, and people would be more likely to travel and holiday closer to home.

The question is: how can we make this rapid transition? Can our current political system, and the imperatives of a deregulated market economy, make this happen very quickly? To be blunt, the answer is no.

Look around for the proof. It is not happening anywhere at the necessary scale and speed. Even in countries that have worked hard to improve energy efficiency, and build renewable-energy capacity and better transport options, the human ability to invent new ways of using energy has worked against these advances: the fast-growing, high-polluting air-travel sector, the air conditioner boom, and the plasma-TV fetish are just three examples. In the West, our conventional mode of politics is short-term, adversarial, and incremental. It is steeped in a culture of compromise that is fearful of deep, quick change—which suggests it is incapable of managing the transition at the necessary speed.

Sharp changes mean disruption, and disrupting business or lifestyle is a political sin. In the developed world, 'politics as usual' places the free-market economy at the heart of its project, and governments, as a matter of political faith, are loath to intervene decisively. Even though Sir Nicholas Stern named global warming as the 'greatest market failure' in history, governments have been ideologically reluctant to act sufficiently to correct this great distortion of the market.

Over the past three decades, just as global warming has slowly become a recognised phenomenon, modern finance capital has extended its hegemony around the globe and, to a remarkable extent, set corporate activity free of national and democratic restraint. Today, the dominant political agenda is for the free market to reign and for capital to be released from government regulation. We hear the mantras endlessly: the public sector is bad, privatisation good; lower taxes, good, government spending, bad. However, as former US labor secretary Robert Reich argued in 2007, in his influential essay 'How Capitalism is Killing Democracy': 'free markets ... have been accompanied by widening inequalities of income and wealth, heightened job insecurity, and environmental hazards such as global warming'. The neo-liberal market economy, without democratic control and with a fetish for monetary growth and 'shareholder value', rather than community, has failed the test of sustainability.

At a book launch in December 2007, Ian Dunlop, former executive for the international oil, gas, and coal industries, said that the crucial issue of the next few decades would be how to 'bring runaway capitalism into alignment with the sustainability of the planet and global society, and indeed with democracy'. He noted that 'the political and corporate structures we have created render us uniquely ill-equipped to handle this emergency,' and that 'perverse [corporate] incentives have led to a paranoia with short-term performance ... organisations previously highly regarded for their long-term thinking have dispensed with that expertise, in the process losing valuable corporate memory'.

He argued that, if we are to ensure long-run sustainability, the rules must change, and he identified three important consequences: genuine sustainable development must become a cornerstone, because conventional growth is untenable; success must be re-defined according to long-term sustainability, not short-term consumption; and markets must be re-designed to enhance local and global 'Commons', a term derived from old English law describing land shared by a village and held 'in common' for the benefit of all. Today, the global 'Commons' refers to all that is central to life, and that no one person or nation should control.

The corporate agenda runs politics, as Robert Reich has articulated:

> Democracy, at its best, enables citizens to debate collectively how the slices of the pie should be divided and to determine which rules apply to private goods and which to public goods. Today, those tasks are increasingly being left to the market ... Democracy has become enfeebled largely because companies, in intensifying competition for global consumers and investors, have invested ever greater sums in lobbying, public relations, and even bribes and kickbacks, seeking laws that give them a competitive advantage over their rivals. The result is an arms race for political influence that is drowning out the voices of average citizens. In the United States, for example, the fights that preoccupy Congress, those that consume weeks or months of congressional staff time, are typically contests between competing companies or industries ... While corporations are increasingly writing their own rules, they are also being entrusted with a kind of social responsibility or morality. Politicians praise companies for acting 'responsibly' or condemn them for not doing so. Yet the purpose of capitalism is to get great deals for consumers and investors. Corporate executives are not authorized by anyone—least of all by their investors—to balance profits against the public good. Nor do they have any expertise in

making such moral calculations. Democracy is supposed to represent the public in drawing such lines. And the message that companies are moral beings with social responsibilities diverts public attention from the task of establishing such laws and rules in the first place.

In short, 'business as usual' practices are no substitute for community-established laws and rules that are created through the state to protect the public good—in the present case, the public good being a healthy planet. Sadly, such a step seems beyond the political process, acting in its usual mode.

Carbon pollution is being turned into a product that, while enormously profitable to its private 'owners', wreaks so much public damage that it threatens to change our planet beyond recognition. Orthodox economic theory demands that the rational course of action would be to place a tax on pollution, so that the cost of the tax would equal the cost of damage being done. This is likely to be very high—Stern says it may be more than US$85 per tonne of carbon dioxide. In fact, if the cost of the marginal damage (destroying the Earth's ecosystems) is beyond value, and of infinite cost, then the abatement price (the amount we should be prepared to pay to stop it) should also be infinite. It seems that this logical conclusion, based on orthodox economics, is not a serious consideration for most people who manage the economy.

To drive the transition to a safe-climate future, greenhouse-gas pollution must be squeezed out of the economic equation. One option is to put an increasing price (tax) on the pollution, so that it becomes more and more economically attractive to use products and processes that do not produce greenhouse gases.

The other option is 'cap and auction', explained in Chapter 20, which is a fancy name for a rationing scheme that sells damage permits in decreasing quantity to polluters, until the economy achieves zero greenhouse-gas emissions.

But there is a problem with both of these solutions. We are addicted to the lifestyle that our high-impact economy allows, which means that — as we have seen with cigarettes and alcohol, which are highly taxed — you can substantially increase the price of greenhouse-gas-intensive products and people will still buy them, because they cannot, or do not, want to go without them, or they are unaware of the low-emission alternatives.

Our addiction to a high-impact lifestyle inbues greenhouse-gas emissions with something called high price inelasticity, which means that an increase in price produces a relatively small drop in demand. In these cases, simple price rises are not an effective, or fair, means for rapidly reducing consumption to zero. As an example, the demand for petrol is highly inelastic, so doubling the price of petrol only reduces demand for petrol by 10 per cent in the short-term, and 40 per cent in the longer term, as people switch to more fuel-efficient cars or other means of transport, or make lifestyle changes. In other words, to reduce the demand for petrol by just 40 per cent, governments would need to double its price, and that is equivalent to a price on carbon emissions of around A$500 per tonne. In the world of 'politics as usual', that is not going to happen.

How, then, can we make the rapid transition happen? Two examples of successful transformations suggest some strategies.

During World War II, after Pearl Harbour, the USA's

military imperatives demanded a rapid conversion of great swathes of economic capacity from civil to military purposes. Within months, car production lines became tank lines. The manufacture of passenger cars ceased for the duration of the war, and new methods to mass-produce military aircraft were devised. Consumer spending was dampened by the selling of 'war bonds' to fund the cost of rapidly expanding military production and to control inflation. Having learned from the devastating experience of profiteering during the First World War, price controls were introduced, and rationing of key goods was mandated as necessary—the main result being a more egalitarian pattern of consumption, especially regarding food. The economy, real wages, and profits all grew, although many civil rights were significantly curtailed.

For a more recent example of successful, rapid transition, we can look to the 'tiger' economies of South Korea, Singapore, Taiwan, Hong Kong, and now China. In each case, national governments and enterprises cooperated in a plan to drive up, or change, the character of their output. In all cases, industrialisation was rapid because domestic demand was held down by state policies that favoured investing in export capacity, savings rates were high, and skills development was emphasised. This is not to glorify these development drives: in the Asian 'tiger' economies there were very significant downsides, including autocratic rule in the service of the development elites, the brutal suppression of labour and democratic rights, the fracturing of traditional rural lives, and massive damage to the environment. Nor are these example given because they exemplify a path to rapid *growth* (which they do), but because they demonstrate the capacity for rapid *transformation*.

What is salient in all these cases is the key role of governments in planning, coordinating, and overseeing the transition—the very opposite of leaving the deregulated market to its devices and going about 'business as usual'. Voluntary measures and aspirational goals will not eliminate greenhouse-gas emissions; they will have to be squeezed out by strong governmental regulatory and investment actions. The particular nature of such a government will depend on the capacity of people to build its democratic character, and to provide national leadership when conventional politics fails to do so. It should not be assumed that strong state intervention requires an autocratic government. If, as a society, we are to engage in a rapid change, it will require the active democratic participation of the population, rather than its passivity.

Even middle-of-the-road climate targets require extraordinary action. The unsafe cap of 2–2.4 degrees that was promoted at the UN Climate Change Conference in Bali in December 2007 requires developed countries to cut emissions to 25–40 per cent below their 1990 levels by 2020. For Australia (where emissions by 2010 will be about 10 per cent higher than they were in 1990), this would require emissions from 2010–2020 to be reduced by 35–50 per cent, or 3.5–5 per cent on average per year.

When the current annual growth in Australian emissions of 1.5–2 per cent is added to the equation, the total turnaround on current practice would be a 5–7 per cent cut in greenhouse-gas emissions each year. By comparison, the best recent record for decreasing the energy intensity of a modern economy is under 3 per cent annually, set by Japan after the 1970s 'oil shock', and achieved, in part, by exporting some energy-intensive industries.

In this context, we find it inconceivable that Australia could play its fair part in meeting even a 2–2.4-degree cap, other than by a planned, rapid transition and economic restructuring. This would necessarily have to be constituted as a climate 'state of emergency' far beyond the capacity of a society operating in its usual modes.

# CHAPTER 23

# What Does an Emergency Look Like?

In recent public discussions about global warming, the language has started to shift from talk of a 'crisis' to one of a 'global emergency'. The popular appeal of Al Gore's film and book *An Inconvenient Truth*, which calls climate change a 'global emergency', has driven much of this shift. In the lead-up to the December 2007 Bali conference, the UN secretary-general also spoke explicitly of a climate emergency.

Using such language is the start of the process of recognising that the underlying reality has become more grave than we had previously realised. The climax of the process may be governments formally declaring a 'state of emergency', at which point we will know a number of things: that the authorities rate the problem as very serious, that priority will be given to resolving the crisis, that we are all in the crisis together, and that, officially, 'business as usual' no longer applies.

The declaration of a state of emergency involves official

recognition that a threat to life and health, property, or the natural environment is sufficiently large that an adequate response will demand a mobilisation of resources beyond the normal functioning of the society. Such threats may be civil or military: they may be natural (like fire, flood, tsunami, or earthquake); political (like war and conflict); biomedical (like infectious disease); or the result of a combination of factors (such as famine or population displacement).

To deal with an unfamiliar emergency, it is often necessary to undertake 'crash programs' to create new capabilities. Iconic examples of such programs have been the Manhattan Project (through which the US developed the nuclear bomb) and the Apollo Program (to get astronauts to the moon). In some cases, the emergency has been so demanding that the whole economy has had to be mobilised to new purposes. Within a year after Pearl Harbour, for example, the US was able to switch from being the world's largest consumer economy to become the world's largest war economy.

In the case of climate change, however, we need to go one step further and change not only *what* the economy produces, but also *how* it produces. Here, the experience of Japan, the Asian tiger economies and, more recently, China is instructive. For example, in two decades, South Korea transformed itself completely from being a poor agricultural economy to a middle-income, world-competitive manufacturing economy. These changes came with very high human and environmental costs, but they demonstrate that programs to transform the organisation of production can be implemented quickly.

Transformational programs can either focus on scaling up existing technologies or processes (to produce a result quickly), or on pursuing fundamental innovation to solve a

| Normal political-paralysis mode | Emergency mode |
| --- | --- |
| Crises are constrained within business-as-usual mode. | Society engages productively with crises, but not in panic mode. |
| Spin, denial, and 'politics as usual' are employed. | The situation is assessed with brutal honesty. |
| No urgent threat is perceived. | Immediate, or looming, threat to life, health, property, or environment is perceived. |
| Problem is not yet serious. | High probability of escalation beyond control if immediate action is not taken. |
| Time of response is not important. | Speed of response is crucial. |
| The crisis is one of many issues. | The crisis is of the highest priority. |
| A labour market is in place. | Emergency project teams are developed, and labour planning is instituted. |
| Budgetary 'restraint' is shown. | All available/necessary resources are devoted to the emergency and, if necessary, governments borrow heavily. |
| Community and markets function as usual. | Non-essential functions and consumption may be curtailed or rationed. |
| A slow rate of change occurs because of systemic inertia. | Rapid transition and scaling up occurs. |
| Market needs dominate response choices and thinking. | Planning, fostering innovation and research take place. |
| Targets and goals are determined by political tradeoffs. | Critical targets and goals are not compromised. |
| There is a culture of compromise. | Failure is not an option. |
| There is a lack of national leadership, and politics is adversarial and incremental. | Bipartisanship and effective leadership are the norm. |

new problem (for example, the Manhattan Project, which set out to build a nuclear bomb, even though the related nuclear industry did not exist and there was virtually no knowledge, at the start, about how to carry out the task). Some transformational programs combine aspects of scaling up and fundamental innovation (for example, the Apollo Program).

All of these very fast, large-scale transformations are characterised by a strong government role in planning, coordinating, and allocating resources, backed by sufficient administrative power to achieve a rapid response that is beyond the capacity of the society's normal functioning.

A state of emergency will likely exhibit many or most of the characteristics listed on the opposite page.

With few exceptions, the present responses to global warming are within the 'normal political-paralysis mode'. Most governments have not been brutally honest with themselves about the new climate data and its consequences, or about the severity and proximity of the consequences if present trends continue. Necessary targets and goals are being severely compromised, while the speed of our response is hopelessly inadequate, and will result in global warming worsening and moving beyond our capacity to construct practical responses. There is neither effective leadership nor bipartisanship.

We are not devoting the necessary resources to solving the problem, whether it is research and innovation, planning for a rapid transition, or scaling up production. Not only has failure become an option; it has also become the norm. On all objective measures the world is going backwards: emissions are rising at an increasing rate, events signalling more dangerous changes in the environment are occurring faster than expected, and positive feedbacks are beginning to kick in.

In short, although it is the greatest threat in human history, global warming is not being treated as an emergency.

This response stands in stark contrast to our behaviour in other emergency situations. In the case of a bush fire, for example, the normal functions of the affected community are suspended, in so far as it is necessary to save life and to devote all available resources — including mobilising them from far away — to fight the fire. The speed of the response is crucial, plans are made in advance, action is centrally coordinated, on-the-ground initiative and committed teamwork are vital, and specialist teams are ready and trained. No effort is spared, people are given leave from their regular jobs, and whole communities support the fire fighters and each other. Resources to fight the fire are not denied because it might 'hurt the economy'.

Yet, in proposing a 'crash program' to curb global warming, the response is often that drastic action is not politically possible — that it will cost too much, damage the economy, waste good capital, or be too disruptive.

Even among many who acknowledge that global warming is an urgent problem, there is a tendency to devalue the predicted impacts. Anyone who talks about living with a 3-degree rise, as some of the climate professionals do, has obviously not come to grips with actual consequences of these figures. In understating the real impacts — and, therefore, the economic damage — the cost of doing nothing, or not enough, is undervalued. At the same time, the heavy cost of action is overstated, especially since many energy-efficiency measures save, rather than cost, money.

It seems that many who are concerned about economic damage are not worried that society as a whole will be worse off by becoming more climate friendly: rather, institutions and individuals who have made themselves dependent on activities that produce large volumes of greenhouse-gas pollution free of charge seem concerned that they will be worse off, and that long-established personal habits and cultural norms will have to change.

The historical evidence, for example, of the emergency mobilisation in the US for the Second World War indicates just how wrong fear-mongering lobbying about 'economic damage' can be. In the period from 1940 to 1945, unemployment in the USA fell from 14.6 per cent to 1.9 per cent, while GNP grew 55 per cent in the five years from 1939. Wages grew 65 per cent over the course of the war, far outstripping inflation, and company profits boomed. This was all at a time when personal consumption was limited by the sale of war bonds, some basic goods and foods were rationed and, at the height of the mobilisation, 40 per cent of the economy was directed towards the war effort.

In 1941, the American economy was still suffering from the effects of the Great Depression, so the switch to the war economy, with its practical need to utilise all available productive capacity, would inevitably produce improved economic statistics. But even if a climate emergency were to be declared at a time of economic health, the tasks are so challenging—building a zero-emissions economy, taking carbon out of the air, and finding the means to cool the planet—that every scrap of productive capacity will be required.

The experience of the Second World War shows that

production and technologies can be switched quickly, and on a huge scale, when there is the need and the will: from a small base of war production in early 1941, the United States was out-producing the combined Axis effort by the beginning of 1943. Merchant shipbuilding grew from a total of only 71 ships for the period from 1930 to 1936 to more than 100 in 1941 alone, and 127,000 military aircraft were produced in the four years from 1941. Output by 1944 was 28 times the rate it had been in 1939.

Lester Brown of the Earth Policy Institute sums up the case for a sustainability emergency:

> The year 1942 witnessed the greatest expansion of industrial output in the nation's history. A sparkplug factory was among the first to switch to the production of machine guns. Soon a manufacturer of stoves was producing lifeboats. A merry-go-round factory was making gun mounts ... The automobile industry was converted to such an extent that from 1942–1944, there were essentially no cars [for commercial sale] produced in the United States.
>
> This mobilization of resources within a matter of months demonstrates that a country and, indeed, the world can restructure the economy quickly if it is convinced of the need to do so. In this mobilization, the scarcest resource of all is time. With climate change, for example, we are fast approaching the point of no return. We cannot reset the clock. Nature is the timekeeper.

Human societies are able to develop emergency methods for handling familiar crises, especially when they are frequently repeated, such as floods, fires, storms, droughts and, in some societies, wars; but we have the greatest trouble with unfamiliar crises, especially if they not yet fully physically apparent.

Now that science, enhanced by its ability to prefigure alternative futures using computer models, has made it clear that a climate disaster is a realistic future, we need to take the crisis seriously. We must treat this future as a preventable fact. We need to start from the assumption that we will not fail in our efforts to prevent this future, and we need to start imagining, and acting out, a whole series of scenarios to prevent climate disaster and to take the world back to the temperature safe-zone. We need to test strategies to see which ones have the highest odds of success. This technique of 'back-casting from success' can be applied to the climate as well as to other issues that need to be tackled under a sustainability emergency.

While the challenge of avoiding climate catastrophe demands action at emergency pace and scale, what has to be done will be very different from responding to cyclones, or wars, or the human consequences of conflicts. Nonetheless, by learning from our past actions at times of emergency, we can prepare for the great task ahead of us all, rather than respond with panic and alarm when it may be too late.

# CHAPTER 24

# The Climate Emergency in Practice

Governments declare states of emergency, or a switch to emergency mode, to signal their profound commitment to solving an urgent problem. Typically, once an emergency has been declared, action takes place at great speed, involving government services and contractors, volunteer organisations, and the active participation of the affected community.

Returning to the safe-climate zone is a more challenging task than is usually faced in an emergency, but the approach to it is similar. Declaring a climate and sustainability emergency is not just a formal measure or an empty political gesture, but an unambiguous reflection of a government's and people's commitment to intense and large-scale action. It identifies the highest priority to which sufficient resources will be applied in order to succeed. Social and economic organisation, and people's everyday lives, will be changed for the duration of the emergency and to the extent necessary to resolve it.

Getting governments to recognise the climate emergency

is a clear and unambiguous goal, but it will face resistance. There will be great pressure for a 'new business-as-usual' politics; governments themselves do not yet fully understand the problem or the solutions; and there is not yet sufficient public understanding of the climate and sustainability crisis, what needs to be done, and why.

One task necessary is to achieve public and government recognition of the scale of the threat, so that serious consideration can be given to the actions necessary to solve the whole problem. Until we see a realistic appraisal, for example, of the Arctic melt and the imminent threats from Greenland and the West Antarctic ice-sheet losses, we will know that political leaders don't yet understand the problem. Once they do understand, and they express the need for goals that will solve the problem, we are on the way to the emergency mode. This step will only occur with broad community education and mobilisation.

Active support will need to be built around both specific goals—for example, large-scale programs for energy efficiency, renewable energy, and restructuring transport—and the formal declaration of a climate emergency.

A street, local area, or organisation, for example, could declare a state-of-sustainability emergency, and could implement an action plan, build up support, and prepare the ground for all levels of government to follow. This would involve initially considering the pros and cons of the emergency idea (for example, by using the scenario in the Appendix), then preparing the full action plan and, finally, executing it. The ultimate goal is to get the government and the whole society to commit to tackling the climate and sutainability emergency.

One way or another, we will get to the emergency mode. The question is whether we can make it happen now, by using our foresight, or whether we have to wait for years until the problem gets so bad that panic flips governments out of their business-as-usual paralysis. How can we act to make the transition to the emergency mode happen sooner, with more chance of success, rather than later, with less chance? How can the ground be prepared to make all the steps feasible?

It is often perceived that the motivation to act strongly on environmental issues comes from a green fringe of society, but this is a simplistic notion. Some of the strongest statements and most effective communications about the climate problem are coming from international figures, such as Al Gore; Yvo de Boer, the UN's lead negotiator for the new global agreement to replace the Kyoto Protocol; long-time sustainability activists such as Lester Brown of the Earth Policy Institute; Amory Lovins of the Rocky Mountains Institute; and many climate-advocacy groups, from the local to the global level.

We now need to substantially restructure the physical economy, in a very short time, but will the mainstream mindset opposed to radical and sudden change be an insurmountable obstacle? We think not, because there are already many people who are very switched on to the problem. In the end, even conservative governments and corporations will figure out that you cannot make money, and grow the economy, on a planet that's not fit to live on. The question is: can we work together to create the rapid transformation of the economy?

John Paul Lederach is a leader in the field of conflict transformation and peace building, and is the author of *Building Peace: sustainable reconciliation in divided societies.*

For a large-scale social challenge in which the whole society needs to move, despite past and current divisions, Lederach argues that change needs to be catalysed at three levels: the population at large; the managerial-cultural-and-innovation elites; and the controlling elite.

In the case of climate, change needs to be occurring at all three levels, and between them. The controlling elite, as a whole, doesn't yet grasp the full seriousness and urgency of the situation, but peer-advocacy from within this level can be very powerful. The other two levels need to direct advocacy to the top, while also making things happen on the ground. Advocacy from members of the managerial elite, including senior administrators and scientists, is important, because they are close to, and provide expertise and management capability to, the controlling elite. Advocacy from the community at large is important, too, because it creates the democratic mandate. While large-scale infrastructure spending and macro-economic management is driven by the elites, initiative and autonomy at the community, workplace, and small-business level is crucially important in changing people's behaviours and activities.

Lederach believes that it is essential to find people who can see the need for change from the perspective of their sector, but who will work across sectors to create cooperation across the fault-lines of conflict.

Because we have not faced a human-caused climate crisis before, there is no complete sustainability emergency package that is ready to be put to use. We need to experiment to find the best way to act, building on our safe-climate message and goals. In creating ideas and imagining the best way to act to make the emergency response a political reality, we can also look around the world to find the most practical and most suitable initiatives.

There are a number of starting points to initiate action on the emergency.

## Mobilising community networks

Transforming politics from conventional to emergency mode will be strongly resisted by those who see a short-term benefit in opposing change. The climate emergency will be resolved only with the active support of the broad community: from community-action groups, unions, and churches to neighbourhoods, schools, and local government; and from political parties to corporate elites. As one expression of this need, in February 2008, participants at a community climate-action conference in Melbourne, Australia, organised by Friends of the Earth and the Sustainable Living Foundation, proposed and set up the Climate Emergency Network (www.climateemergencynetwork.org). The most active local community climate groups across Melbourne and regional Victoria now support this project, which will initiate community mobilisation for strong and effective action on climate change and the declaration of a sustainability emergency. Building democratic participation is one key to achieving emergency action.

## Building deliberative democracy

People and organisations need the opportunity to look at all the issues in depth and over time, in a supportive social environment. Because the worst climate impacts are still to come, full-strength, effective action must arise from an act of informed imagination, and we need to help large numbers of people do this. The techniques for doing so fall under the concept of deliberative democracy.

The Victorian Women's Trust provided an inspiring example of deliberative democracy with their Watermark Australia project. The project engaged 2000 people, across Australia, in 200 discussion groups that met for two periods of three to five months each. Discussion groups were the core of the program because many people learn better and become more motivated to take action when involved in a group. In the first period, the groups discussed how the water system works, or does not work, in Australia. They looked at a range of issues—including what is happening to rainfall, how water is used and could be better used, how society and the economy interact with water issues, and what water's environmental role is. In the second period, participants developed ideas for action on the key issues.

The process included a two-way exchange between program participants and water-issue experts. The results were drawn together into the book *Our Water Mark*, which had more than 37,000 copies distributed across Australia.

A similar in-depth engagement of a significant proportion of the community would be an asset in developing responses to global warming. Repeated interaction of this sort would ensure that the community is fully engaged, that governments gain a specific democratic mandate for emergency-scale action, and that politics becomes a process that engages people constructively with tough issues that need strong responses.

Deliberative democracy can build awareness of the sustainability emergency and create the political space that governments need to act. Far from the emergency causing democracy to suffer, it could be a decisive factor in making the democratic process work more effectively.

## Figuring out how it can work

Because no region has ever declared, or tackled, a sustainability and climate emergency, advocates and organisations are uncertain about what is involved and how to best proceed. In part, this can be overcome through forming, or strengthening, existing non-government research bodies, or think-tanks, that are open to expertise from across society. They would study all the detailed problems related to the emergency and make their insights available to community groups, businesses, and unions, as well as provide advice to advocates, lobbyists, and government. Their investigations could include the administration of the emergency, proposed legislation, action plans, solutions to the political obstacles, economic policy and management, and how to drive innovation. Research would aim for the very best possible model to run the sustainability emergency — one capable of delivering a fast return to a safe climate while enhancing democracy by treating all parties fairly and with respect.

## Experiencing the climate options

While we need to understand the implications of dangerous climate change rationally in order to think about the possible solutions, our ability to act decisively on the sustainability emergency will be assisted by a capacity to also 'feel and see' it — to empathise with where our planet is headed, and to be able to think about how different it might be.

One way to do this is to 'virtually' experience climate catastrophe and alternative futures. Books such as Mark

Lynas' *Six Degrees* and Fred Pearce's *With Speed and Violence* are important. Science fiction books like Kim Stanley Robinson's climate-change trilogy, *Forty Signs of Rain*, *Fifty Degrees Below*, and *Sixty Days and Counting* help us to imagine living in dangerous climates that experience abrupt change.

If this assisted imagining is to be helpful, it must also enable us see and feel what it would be like to take action to rescue the planet now, while we have the best chance of success, and well before widespread catastrophe hits or becomes unavoidable. We need stories not just of heroes living in the future, and making desperate, perhaps futile, last-ditch attempts to head off a disaster in the face of imminent doom, but of a world in which climate turnaround is achieved, despite distractions, inertia, and vested interests in the here-and-now — an interesting challenge for sustainability strategists, writers, and movie-makers.

## Creating a radical innovation program

Transforming society to achieve a safe climate is a complex task requiring a high level of innovation. A trial of a landmark approach, supported by five government ministries, took place in the Netherlands during the 1990s, and the results were published in the book *Sustainable Technology Development*.

The Dutch Sustainable Technology Development Program found that a 95–98 per cent improvement in eco-efficiency was required to reduce the environmental impact of production, taking into account the current, too-large levels of damage and waste, and the projected population and economic growth. It was realised that normal innovation, dominated by incremental change to existing technologies, would not be able to deliver the required level of improvement. The solution was to go back to basics and identify the human and environmental needs that had to be met and then, from first principles, invent new technologies that could meet all these needs efficiently. A series of case studies showed how this approach could be applied successfully to a range of human needs, including the demand for high-quality protein, water services, clothes-cleaning services, provision of chemical feedstock, and mobility. The most dramatic success was a redesign of ways of providing high-quality food protein, which saw a 99 per cent improvement in eco-efficiency.

Faced with a climate emergency, we need to be innovative to deliver eco-efficient technologies that change how we live and how we produce goods. One stream would have to involve a 'crash program' to deliver large results quickly; another stream would include fundamental research to find increasingly efficient and effective ways to meet human needs and the environmental needs of 'all people, all species, and all generations'.

The 'crash program' stream could learn from the Apollo Program experience and the rapid industrial transformations of the tiger economies.

The 'fundamental innovation' stream could help the crash programs avoid options that have long passed the point at which they can make a positive contribution to a safe-climate program. An example is the idea of replacing coal with natural gas (which causes continuing high levels of carbon emissions) as a long-term generator of electricity, when what we need to do is to drive down emissions to zero.

## Developing a public computer-modelling agency

We now have powers that run far ahead of our capacity for

foresight. When figuring out solutions to restore a safe climate, it is very hard to work out the unintended consequences or the exact impact of various plans.

Climate scientists have sophisticated models to help their research, but these are not available to the general public. It is not possible to find answers to many pressing issues unless you have enough money to commission research, or are lucky enough to find a research institution that has already answered your question.

In preparing this book, there were many questions to which we could not find a ready answer. How much would the complete loss of the Arctic sea-ice raise regional and global temperatures? We learned that no specific modelling has been done on this question, even though complete sea-ice loss now seems imminent. How much greenhouse gas would need to be taken out of the air to trigger sufficient cooling to get the summer Arctic sea-ice back? What would a safe-climate emissions scenario look like? What is our capacity to produce biochar, using existing waste biomass?

Establishing a public utility for modelling changes in climate, the environment, and the economy, at least in outline, would allow us to answer such questions. We could find the answers to politically inconvenient questions that government, universities, or other institutions may be tempted to self-censor or avoid. The impacts of alternative policies, programs, or products could be explored. Models need to be accessible to everyone so that people and organisations can test their policy or development ideas. Arguably, the modelling capacity should be available free, or so cheaply that everyone can make use of the facility, which would allow public discussion to be concrete and proposals to be more thoroughly tested.

We do not pretend to have a detailed roadmap for tackling the sustainability emergency. The ideas presented here are simply an outline of plans and actions that are necessary and workable.

# CHAPTER 25

# The Safe-Climate Economy

We have gone too far. The planet is already too hot. There is too much carbon dioxide in the air, the Arctic will soon be ice-free in summer, and the world is committed to more than 3 degrees warming if it continues along the present energy path.

We are not standing on the threshold of dangerous climate change; we passed through that doorway decades ago. Will we take action, at great speed, to rescue ourselves and the other species with whom we share this planet, or will inadequate action condemn the living Earth to catastrophe?

The rescue will be no small task. We must cool our planet by changing our energy systems and the way we move, work, and produce. This requires the redirection and reinvention of much of the economy in the shortest possible time. If we value life, the time has come to incorporate climate-change objectives into the structure of economic management.

In 1929, much of the world was plunged into economic depression and mass unemployment. In previous economic collapses, unemployment was treated as a personal tragedy for the individual, because the downturn was accepted as a natural feature of the economy. During the Great Depression, unemployment became an issue to be consciously dealt with by economic management. New economic theories (most famously, those of John Maynard Keynes) and the rising power of organised labour demanded action to restore jobs and economic balance.

While the Depression gave birth to macroeconomic management, the Second World War drove the development of many of the tools. There was an imperative to harness the entire industrial economy to the war effort. Systems of national accounts and indicators such as Gross Domestic Product were refined, and methods for enlarging and steering the productive capacity of the economy were created.

The climate and sustainability emergency presents a remarkably similar challenge. The behaviour of a world economy that has been unconcerned with environmental costs is the principal cause of the emergency; but the solution lies in harnessing its productive capacity to build climate-safe infrastructure and macroeconomic governance which ensures that such severe problems never arise again.

The relative prices of raw materials, energy, environmental capacity, and ecosystem services must be set so that, socially and privately, the most profitable action coincides with protecting the environment and conserving resources. Two important tools are eco-taxation and rationing, the latter setting a declining limit on environmentally adverse economic activities by auctioning permits. The revenue raised can be used to make changes in infrastructure, industry sectors, and regional economies to deliver goods and services in new ways. This is the domain of industry policy and regional-development policy. New policies are also needed to build skills and capabilities, and to foster innovation and strategic management.

Plans to reduce greenhouse emissions must be comprehensive. We could virtually eliminate climate pollution from most aspects of our lives, but if just one sector—transportation—were to be overlooked, our efforts would be undone.

Air travel as a sector, for example, is the fastest-growing producer of global greenhouse emissions. Aircraft emissions from high in the atmosphere have an effect 2.7 times as great as the same carbon dioxide pollution at ground level. There are few readily available low-impact fuel substitutes. In Australia, taking into account the projected increases, total air-travel emissions by 2020 will have an impact equivalent to two-thirds of a tonne of carbon per person. Air travel alone will be enough to blow a sharply declining carbon budget.

The total quantity of all greenhouse emissions would be best controlled by rationing, rather than by standard eco-taxation. Existing 'cap and trade' schemes are generally deficient: they include only some emissions, they give away permits, they legitimise rorts, and they fail to deal with cross-border problems.

There are, however, some good rationing models that have been proposed. An approach that has been considered by the British government, and that seems well suited to the demands of the sustainability emergency, is the introduction of personal carbon allowances (or rations) to guarantee that the national greenhouse emissions budget is achieved.

The system could work in an Australian context as part of a safe-climate strategy. An authority independent of government, like the Reserve Bank, would set up a national greenhouse-emissions budget each year. The amount of emissions would be decreased each year, through a series of downward steps (to zero), in accordance with a rapid transition plan. Because households (in Australia) are directly responsible for about one-quarter of emissions (generated principally by household energy use and private travel), one-quarter of the carbon budget would be made available free of charge to each citizen as an equal 'carbon credit' (or ration), via an electronic swipe 'carbon card'. The card would be used to draw on an individual carbon-credit balance each time household gas and electricity, petrol, and air tickets were paid for. Unused credits could also be sold. For the energy embedded in purchased commodities, such as food and personal services, the carbon ration would already have been for paid by the manufacturer, and its cost would be built into the consumer price. If a person lacked the greenhouse-emissions credits to cover a purchase, or they were an overseas visitor without an entitlement to emissions credits, they could buy credit at the point of sale.

The balance of three-quarters of the national emissions budget would be auctioned to business and government in an emissions-credit market, where the price of emissions would rise over time as the quantity was progressively reduced. Businesses or individuals would also be able to sell excess emission credits. Because individuals and businesses would be able to trade their credits within the overall limit set by the national greenhouse-emission target, there would be a financial incentive to make a rapid switch to low-emissions technologies. If a new technology required less of an individual ration compared to the technology it replaced, it would be more attractive, so businesses would have an incentive to make long-term low- and zero-emissions investment decisions.

Rationing is feasible and was used very effectively during the Second World War and for some years afterwards. Studies in Britain showed that war rationing was accepted because it was seen as both necessary and fair. There was a booming black market, in which rations were bought and sold informally, because goods were in short supply and ration cards could not be officially traded. But in response to shortages of some goods—because war needs were a higher priority than consumer demand, or because precarious merchant shipping could not carry the volume of imports required for normal market trading—the population accepted the argument that, for basic necessities such as fuel, some foods, and clothing, it was fairer to get an equal ration than to restrict demand by raising the price. This contrasted with the very unhappy experience in the First World War, in which the free market had operated and there was rampant profiteering in the face of shortages.

British feasibility studies suggest that perceived limitations of a carbon rationing system could be resolved; for example, concern about the capacity to efficiently administer and track people's carbon allowances is unfounded. The transaction costs of using a personal carbon 'smart card' would not be overwhelming and, in practice, would be less demanding than systems like Australia's Medicare health care scheme.

We are already seeing the rising cost of energy, water, and food to households and consumers, for reasons including the increasing price of oil, international competition for secure energy supplies, and rising energy costs. These rises also reflect, in part, the higher cost of water, which is used in large quantities in coal-fired generators. One manifestation of the sustainability crisis that we can see today is increasing fuel poverty in lower-income households. Carbon rationing could exacerbate fuel poverty, but all measures that put a price directly on carbon, such as taxes, create this problem. Studies in the UK show that carbon allowances would be more progressive than a carbon tax. Even if the revenues from a carbon tax were recycled through the tax system as effectively as possible through targeted increases in benefits to low-income households, carbon rationing would still produce a fairer outcome.

Either way, there is a need for government to mandate and provide resources for upgrading the energy efficiency of domestic and commercial buildings, and goods and services, so that people on low incomes do not face unmanageable costs. Such programs are well established in countries such as Germany and the United Kingdom.

Personal carbon rationing appears more equitable than the alternatives. Because rationing works by imposing quantity restrictions at the outset, rather than by raising prices, it does not in itself increase the price of the household and personal energy consumed—provided that society takes steps to create an economy, including affordable goods and services, that does not require the emission of greenhouse gases. Rationing is also fairer than increasing taxes, because personal greenhouse-emission allowances provide free entitlements to individuals and only impose financial penalties on those who go above their entitlement, while providing an income supplement to those who use less than their entitlement. In general, people on low incomes use less energy and emit less carbon dioxide than average (particularly if personal air travel is included), while wealthier people consume more resources and are, therefore, responsible for a greater-than-average level of emissions. Wealthier people will need, on average, to buy allowances from poorer people, who are likely to use less than their ration.

Rationing offers a number of other benefits. It is egalitarian, in that everyone gets an equal, free carbon allowance; it allows people to make choices and to create a personal carbon budget, which is more empowering than simply watching prices automatically go up; it encourages positive behavioural changes, in the knowledge that others, including businesses and the government, are also acting within the scheme; it helps address our depleting energy resources; and it is more effective in reducing emissions when targets are strong.

David Miliband, then British environment secretary (and now foreign secretary), told an audience in 2006 that 'the challenge we face is not about the science or the economic … it is about politics'. He said that carbon rationing can 'limit the carbon emissions by end users, based on the science, and then use financial incentives to drive efficiency and innovation'.

It does not require much imagination to understand that the corporate 'big end of town' may see the idea of rationing as a direct challenge to their world and to their idea of a free market. They express the fear that strong action to make a safe climate possible will destroy the economic-growth machine. There are a number of responses to this.

Their fear makes no distinction between the fate of an individual corporation and the fate of the economy at large. A safe-climate economy will differ greatly from the current one, and firms that are not adaptable, and that remain in the old economy mode, may fail. The economist Joseph Schumpeter famously described the modern industrial economy as unleashing gales of 'creative destruction', with waves of innovations sweeping away the innovations of an earlier time. While technologies, products, and organisations might wax and wane, the vibrancy of the economy, as a whole, can be maintained.

For several decades to come, the challenge that we face will be to satisfy the basic human needs of more than six billion people while, at the same time, carrying out the most profound rebuilding of the world economy since the beginning of the industrial revolution. There is also the challenge to take at least 200 billion tonnes of excess carbon dioxide out of the air, and to help the world cool down in other ways while efforts to cut emissions to zero and draw down carbon dioxide take effect.

Far from causing economic collapse, the challenge of this fundamental change is more likely to be in expanding productive capacity. This will require high levels of economic output and employment, in the service of achieving full-strength ecological sustainability. While those companies that choose not to contribute to this transition will lose out, the economy as a whole will prosper.

The 'new business-as-usual' approach, in failing to understand the severity of the problem and the depth of action required, implements half-measures that will slow, but not stop, the onset of climate catastrophe, and that may worsen the situation. These half-measures usually build on the core competences of our materials-intensive economy, allowing companies to stick closely to what they already know: how to make more and more material goods while using up more and more 'ecological space'.

Biofuels are a good example of one of these measures. This approach is not ecologically sustainable and cannot deliver the results we need over the medium to long term. As long as the dominant economic players try to solve the climate emergency by using the 'new business-as-usual' paradigm, there will, necessarily, be conflict between them and those adopting a truly sustainable approach.

Can these forces be reconciled? In both approaches, economic growth is a key issue. To the 'new business-as-usual' side, it is a sign of success; to the other side, it is a measure of the rate of ecological destruction. But neither side is necessarily correct.

Setting aside the question of whether economic growth *should* be a key goal, are there conditions in which economic growth could go on indefinitely, while still being ecologically sustainable? The answer is yes, in the following circumstances:

- when the population is stable and not too large;
- when the economy operates with a non-growing (effectively fixed) quantity of materials, energy, area of land, and water environment;
- when the quantity of materials, energy, and ecological space is reduced dramatically, compared to the present, in order to provide for the restoration and maintenance of enough quality habitat for all other species and ecosystem services;
- when resources are used with the utmost material efficiency;
- when energy is renewable, and materials are recycled as close as possible to 100 per cent; and

- when chemical intrusions into the environment are reduced to safe levels and do not systematically increase.

These are some of the key requirements for achieving ecological sustainability. Importantly, when the focus is on the *qualitative* growth of service-value, rather than on pumping out more and more material production, economic growth could continue indefinitely, without clashing with ecological sustainability. Indeed, this is the model that evolution has worked on for at least the last several hundred million years.

Economic growth is said to be central to economic development, but what actually drives the latter is innovation and invention. Economic growth *per se* is a red herring.

The more committed a society is to economic growth, the more it must also be committed to fully eliminating its consequent negative environmental impact. A growth economy that ignores the needs for ecological and social sustainability will, in time, destroy its own foundations and, ultimately, collapse.

For the duration of the sustainability emergency, economic innovators and advocates for ecological sustainability must work together to meet human needs and build a path away from climate catastrophe. This essential collaboration must rapidly take a radical new direction, away from the 'new business-as-usual' economy and to an ecologically and socially sustainable safe-climate economy.

# CHAPTER 26

# In the End

If we are serious about creating a safe climate quickly, how much of the world's economic capacity should be devoted to making such a rapid transition? Some economic modellers and policy-makers have been bickering over tenths of a per cent, and fantasising that the world might be able to avoid dangerous climate change while 99 per cent of the economy continues as before.

We can't emphasise strongly enough our view that we must all devote as much of the world's economic capacity as is necessary, as quickly as possible, to this climate emergency. If we do not do enough, and do not do it fast enough, we are likely to create a world in which far fewer species, and a lot less people, will survive. It makes no sense to give high priority to producing yet more 'cream on the cake' when the very viability of the planet, as a life-support system, is at stake.

We are close to blowing the system, as many leading figures are now saying with increasing urgency. At the 2007 Bali conference, UN chief climate negotiator Yvo de Boer said that reducing emissions by 25–40 per cent by 2020 would cap global warming to 2 degrees, but that this could still result in 'catastrophic environmental damage'.

It is now or never for truly radical action and heroic leadership. During the last global mobilisation, the Second World War, more than 30 per cent, and in some cases more than half, of the economy was devoted to military expenditure. As the table below demonstrates, that economic transformation was achieved in a few years:

**Military burden 1939–1944**
*(military outlays as a percentage of national income)*

|         | 1939 | 1940 | 1941 | 1942 | 1943 | 1944 |
|---------|------|------|------|------|------|------|
| USA     | 1    | 2    | 11   | 31   | 42   | 42   |
| UK      | 15   | 44   | 53   | 52   | 55   | 53   |
| Germany | 23   | 40   | 52   | 64   | 70   | –    |
| Japan   | 22   | 22   | 27   | 33   | 43   | 76   |

Source: Harrison, M. (2000) *The Economics of The Second World War: six great powers in international comparison*, Cambridge University Press

As a rough estimate, A$300–$400 billion invested in renewable energy and energy efficiency in Australia would allow the nation to close every coal-fired electricity generator; it would transform key industries, and the rail and transport system, and provide a just transition for those who might be economically displaced by the changes. Much of that investment in energy efficiency would also be repaid, over time, in energy cost-savings. An investment of that size would be just 3–4 per cent of total economic production for 10 years, minus the energy savings, which is minuscule compared to the Australian war effort. Is it beyond a developed country's ability to identify 3–4 per cent of total personal consumption, government expenditure, and corporate activity that could reasonably be re-directed to this necessary task? It seems a very cheap up-front price to pay; and the nation would reap the rewards of this investment forever.

Some will argue that the cost would be even lower, as the McKinsey & Company report on Australia suggested; however, that calculation was based on national emission cuts of 60 per cent by 2050. Dealing with the real emergency, and the scientific imperatives it has unleashed, requires a much steeper and more rapid emissions-reduction curve, along with the introduction of cooling mechanisms. It also requires us to assist poorer countries which are less responsible for the predicament that the world faces, and less able to respond.

One objection to this vision of a rapid transition to a post-carbon economy is that some power-generating companies may go out of business, undermining one of the great institutions of modern life: shareholder value. Through superannuation funds, many of us are shareholders in those electricity generators. But, in reality, some power plants are reaching the end of their anticipated life-spans and have already been depreciated. Others that aren't at this point may be compensated. Addressing this effect seems to present a far smaller challenge compared to letting coal-fired power stations, and other users of fossil fuels, wreck the climate. Contrary to a preoccupation with 'shareholder value', we suspect that most citizens would think the greatest 'value' would be a viable future for our planet, our lives, and our children; in other words, that they would welcome an end to the fossil-fuel industries and, in their place, the development of sustainable industries. The chorus of respectable voices, including that of Al Gore, calling for an end to era of coal-fired power generators is growing rapidly.

Really, our main problem is political inertia, not the cost to the economy. It will cost an estimated US$130 billion to ensure that all Indian households enjoy access to electricity by 2030. Let us say that this cost would be doubled if it came from renewable sources. That would be $20 billion a year for 15 years, or around 3 per cent of the annual US military and intelligence budget (including Iraq and Afghanistan), which was US$700 billion in 2007. Just two years of US spending in Iraq and Afghanistan would more than pay the whole bill. So it is not a question of having the money; rather, it is a matter of the choice that we, and our governments, make about where to spend it.

Wherever you look, the story is the same. It is estimated that it might cost an additional US$30 billion per annum to put in place safe-climate power supplies in countries outside the OECD. This would amount to less than 0.1 per cent of the total annual production within OECD countries. Compare that to a world war, during which antagonists devote a third of their economy, and often more, to military spending.

Yet, while every nation on Earth is threatened by catastrophic global warming, most governments are still refusing to act with sufficient speed or financial commitment, exhibiting little courage, foresight, or capacity.

Many of us — in business and at work, in climate-action groups, in the not-for-profit sector, and in political parties — know in our hearts that, in avoiding tackling climate change, these governments are showing poor leadership, and that the solutions which currently dominate national and global forums are inadequate. Sometimes, though, we dare to imagine that we could mobilise, on a great national and international scale, a very rapid transition to a safe-climate, post-fossil fuel, sustainable way of living.

We now need to imagine such a course of action, because a sustainability emergency is not a radical idea. It has become necessary to save our future.

# A *Climate Code Red* Scenario

An effective way to introduce *Climate Code Red* to organisations is through a strategic 'planning scenario' that allows people to examine the problems in-depth without prior commitment.

Scenario planning is a method used to examine alternative futures, ranging from those considered the most likely to less likely but significant possibilities. Exploring scenarios helps organisations to respond to change and unexpected events; for example, oil companies might run scenarios that examine the effect on their business of war in a large oil-producing country, or the implications for sales of a global recession.

Scenario planning can identify features of the future that the organisation would like to help bring about, or new activities that position it well across a wide range of futures. It can also help to prepare contingency plans for less likely events or trends.

Scenarios often distinguish between low and high threats, and partial and full responses. A plausible range of responses to a high-level climate threat could be:

- failed cooperation (everyone for themselves), in which no real agreement is reached between the most influential players;
- the agreement of critical parties on partial measures; and
- the timely agreement of critical parties on safe-climate measures (full-strength measures that solve the problem).

A single scenario does not account for all possible futures, but paints a picture that may reasonably occur, based on a set of specific choices or external events. Strategic planners can then explore the dynamics, consequences, problems, and opportunities that arise within the scenario, and examine the merit of different courses of action.

Two key features differentiate our *Climate Code Red* worldview from other climate-change responses: it considers the climate threat to be larger, and more urgent, than most analysts suggest; and it proposes a full-strength response to achieve a return to a safe climate, rather than merely a slower onset of catastrophe.

Here is one of several scenarios that could be drawn from *Climate Code Red*.

## The scenario trigger

One summer, sometime between 2009 and 2013, all the Arctic Ocean sea-ice melts. It then reforms in winter, and completely melts again each summer thereafter, initiating warming that will, in time, cause a rise in regional temperatures of 5 degrees, and a global rise of 0.3 degrees, as light-reflective ice is replace by heat-absorbing dark seas. This causes an accelerated melting of the Greenland ice sheet—which is predicted, along with other factors, to increase global sea levels by up to 5 metres by 2100.

Sea-level rise is exponential, starting slowly and quickening towards the end of the century, so that early solutions are very valuable. The increasing Arctic temperature also accelerates the melting of the permafrost soils, releasing additional large amounts of greenhouse gases, particularly in the second half of the century.

Unless effective action is taken to tackle global warming, the current climate-trajectory already commits the Earth, over the longer term, to a temperature increase, compared to the pre-industrial temperature, of at least 3 degrees. This trajectory includes the fact that:

- a rise of 0.8 degrees has already been experienced;
- a rise of 0.5 degrees is predicted to occur over the next two decades; and
- further warming (which, while taking some time to reach full measure, is already having an effect) is being caused by:
    - the loss of the Arctic sea-ice (0.3 degrees);
    - growing emissions from melting permafrost (0.7 degrees); and
    - air-temperature rises delayed by the warming of the oceans (0.7 degrees).

To these could be added warming from many other causes (including the increased frequency of forest fires), and the declining effectiveness of natural carbon sinks.

If this potential temperature increase were to be realised,

there would be catastrophic results for people and for other species.

## The possibility

Eminent climate scientists have drawn attention to the increasing impact on the climate of positive feedbacks, which reinforce and amplify human-caused global warming in the natural world to produce much greater warming. But it is recognised that the reverse is also true: human actions that result in sustained climate cooling will trigger natural processes that drive further cooling, so that reaching a lower, safe global temperature is made easier. Scientists also suggest that the Arctic ice can be restored to its normal physical extent (as experienced during the past 10,000 years) relatively quickly—over perhaps a couple of decades—if there is modest cooling that returns the polar north to pre-1980 temperatures.

## Scenario values

This scenario is premised on:

- achieving a safe climate in the interests of all people, all species, and all generations;
- a low acceptance of risk, as found in best-practice engineering; and
- applying the principle of 'double-practicality'—action must happen in the real world and that must fully solve the problem—and the attitude that 'failure is not an option'.

## The conundrum

At a practical and physical level, the scenario is based on a conclusion that has been drawn from the science that a safe-climate future is not possible if the Arctic icecap is permanently absent during the northern summer.

It is estimated that, to restore the Arctic ice, the global temperature needs to drop by at least 0.3 degrees from the 2008 level, and that the long-term level of greenhouse gases in the air needs to be in the range of 300–325 per million parts of carbon dioxide. To achieve this cooling, we need to set a greenhouse-gas emissions target at zero, and take other significant measures as well. In total, we need to draw about 200 billion tonnes of carbon out of the atmosphere, in order to reduce the heating effect of excess carbon dioxide already in the air, which will in turn fully restore the Arctic icecap in summer.

However, as we cut carbon dioxide emissions, we also reduce the release of aerosols that accompany fossil-fuel combustion. Aerosols, on average, act as a cooling agent in the atmosphere, but are washed out of the lower atmosphere within two weeks; on the other hand, carbon dioxide is only removed from the atmosphere very slowly, and acts as a warming agent for hundreds of years. So if we stop burning fossil fuels, there will be a once-off temperature increase of at least 0.7 degrees, because the accumulated effect of past carbon emissions continues at the same time as the cooling effect of recently emitted aerosols is rapidly lost.

This spike in temperature could be partially, but not fully, offset if a major effort is made to reduce the emission of short-lived greenhouse gases, such as methane and releases of black carbon.

The removal of atmospheric carbon dioxide can be accomplished by growing biomass, converting it to biochar (which is largely carbon), and sequestering it in agricultural soils; or it can be accomplished by fully combusting the biomass, and sequestering the carbon dioxide in geological structures. Simply regrowing forests as a way of storing carbon will not produce enough cooling. Even large-scale biochar production, or combustion and sequestration methods, could take a large number of decades — perhaps as long as a hundred years or more — to initiate effective cooling. As more of Greenland melts and rising temperatures prompt other warming effects, even a transition of decades will prove to be too slow.

## What action should we take?

At the time of writing, there were no scientific estimates in the peer-reviewed literature stating exactly how quickly the industrial restructuring and climate-system-change process would take to achieve a safe climate. The *Climate Code Red* scenario assumes that the industrial transformation needs to be as fast as can be made possible, for the following reasons:

- the planet is already too hot, as is now particularly evident in the Arctic;
- high rates of temperature increase will tear apart natural ecosystems;
- extreme weather events and climate changes are already debilitating many people and nations; and
- there are many unpredictable possibilities that could arise as a result of current greenhouse-gas levels and near-term temperatures: they may destabilise

the tropical rainforests and cause their collapse after severe fire; they may destabilise the West Antarctic ice sheet and lead to a catastrophic release of ice into the oceans; and they may cause natural-system warming feedbacks so strong that human efforts to orchestrate cooling will no longer countervail the warming forces.

The fastest restructure of a modern economy occurred during the Second World War, and it seems likely that, with full mobilisation, the industrial restructuring that is needed could be completed in about a decade. The *Climate Code Red* scenario assumes that society will, in due course, attempt to complete such an economic restructuring. This will stop greenhouse-gas levels in the air from rising, and will initiate the accelerated removal of excess carbon dioxide from the air; however, in view of the damage already being done by climate change, the beneficial effects of this transformation will almost certainly be too slow in coming.

Two key issues arise in this scenario:

- we have to stop emitting greenhouse gases quickly, because the more we emit, the bigger the eventual temperature rise will be; however, in cutting greenhouse gases that are generated by combusting fuels, the aerosol effect will cause a serious short-term temperature rise; and
- we must stop the temperature from rising too fast, or too far, and we cannot allow high temperatures to persist for too long, otherwise too much damage will be done.

In light of these issues, we will need to apply additional strategies. Direct cooling strategies that work by increasing the reflectivity of the Earth are most likely needed. These include taking actions that increase the cover of highly reflective cloud (for example, by boosting plankton growth in the oceans or by re-establishing forests), or injecting aerosols into the upper atmosphere (where they are not washed out by rain), which will also boost reflectivity.

In the last 50 years, humans have been unintentionally geo-engineering the Earth on a huge scale by releasing into the air large quantities of greenhouse gases and partially countervailing aerosols. Enormous care will be needed to determine the extent to which direct cooling is needed, and to design and select direct cooling methods that can produce clear-cut environmental benefits. The use of temporary intentional geo-engineering for cooling purposes must not be used as an excuse to prolong the release of carbon dioxide.

## Economic and political consequences

Under the *Climate Code Red* scenario, there are three enormous tasks that will absorb a sizeable portion of the global economy's productive capacity, particularly during the decade or so in which the bulk of the physical restructuring takes place: the global move to zero greenhouse-gas emissions in as short a period as is environmentally safe, the drawing down of many billions of tonnes of carbon from the air over the fewest possible number of decades, and the direct cooling of the Earth for as long as necessary. Directing a necessarily large part of the economy to the task of creating a safe climate is not seen as being possible under normal political conditions.

The physical success of the scenario depends on sufficient

action being taken by nations that produce most of the emissions and that have the economic and physical capacity to contribute to the drawdown of carbon dioxide and to direct cooling. To make this commitment socially possible, we assume that nations will conclude that they need to go into emergency mode; but the type of emergency action needed is of an unprecedented form.

The dynamics of the climate-change challenge are different from those of the Second World War, during which the threat was palpable from the beginning. While many societies are now feeling significant climate impacts, the largest effects of the current level of greenhouse gases will be felt several decades into the future, and so the degree of action that is necessary now is much stronger than is justified by current impacts alone.

All countries, no matter what their political system, whether liberal-democratic or not, will struggle to achieve the needed change unless they engage their communities in a deliberative process to learn about the climate-change issue, and help them to reach a genuine understanding of the severity of the problem and the necessity for urgent action on a huge scale.

### Questions to ask about the *Climate Code Red* scenario

1. How valid is the assessment of the climate science? Could the threat be as serious as is argued in *Climate Code Red*?

2. Is it possible to avoid or reverse dangerous tipping points such as Greenland ice-sheet disintegration, sizeable permafrost carbon emissions, or the catastrophic conversion of rainforests (that is, the conversion of the Amazon to savannah grassland) if the Arctic remains free of sea-ice in summer?

3. Who could give a well-informed assessment of the science arguments in *Climate Code Red*, given that much of the science relied on goes beyond the current IPCC consensus?

4. If the scientific conclusions of *Climate Code Red* are reasonable, are the proposed responses appropriate? For example:

- How necessary are the ethics of creating policy for the benefit of 'all people, all species, and all generations', and the ethics of being risk averse?
- Is the idea of going for a 'safe climate' rather than just avoiding dangerous (catastrophic) climate change a sound response?
- How accurate is the idea that the required solutions go so far beyond business-as-usual that a sufficient response is possible only by establishing the problems as an emergency?

5. How might the problems of the aerosol conundrum, or of establishing a cooling mechanism at a sufficient pace and scale, be resolved?

6. How could the detailed action specifications of the *Climate Code Red* scenario be improved?

7. Will your organisation take a proactive position on a 'safe-climate' future, or will it act neutrally, or in opposition, to such a future?

8. Will your organisation prepare itself to prosper, or function well, in a 'safe-climate' future?

Note: Updated versions of the scenario will be available at www.climatecodered.net

# Notes

## Introduction: A Lot More Trouble

Ban Ki-Moon's statement was reported by ABC News, 'UN chief says global warming is "an emergency"', 11 November 2007. Richard Alley's '100 years ahead of schedule' was reported by P. Spotts, 'Little time to avoid big thaw, scientists warn', *Christian Science Monitor*, 24 March 2006. Jay Zwally's canary metaphor was reported by S. Borenstein, 'Arctic sea ice gone in summer within five years?', Associated Press, 12 December 2007. Robert Corell's statement was reported by I. Hilton, 'Greenland is now a country fit for broccoli growers', *The Guardian*, 14 September 2007.

North pole tipping point passed: M. Inman, 'Global warming "tipping points" reached, scientist says', *National Geographic News*, 14 December 2007.

Nicolas Stern's 3-degree goal is analysed in Chapter 11.

## Chapter 1: Losing the Arctic Sea-Ice

Martin Parry's statement was reported by D. Adam, 'How climate change will affect the world', *The Guardian*,

19 September 2007. The February 2007 IPCC report is *Climate Change 2007: the physical sciences basis, Working group I report* (IPCC: Geneva). Tore Furevik's presentation is 'Feedbacks in the climate system and implications for future climate projections' www.norway.org/NR/rdonlyres/3F179CEC-67E4-4512-8229-701B48B5E54E/36279/fureviktore1.pdf.

Marika Holland's comment was reported by J. Amos, 'Arctic sea ice "faces rapid melt"', BBC News, 12 December 2006. The March and May 2007 studies are M. C. Serreze, M. M. Holland et al. (2007) 'Perspectives on the Arctic's shrinking sea ice cover', *Science* 315: 1533–36; and J. Stroeve, M. M. Holland et al. (2007) 'Arctic sea ice decline: Faster than forecast?', *Geophysical Research Letters* 34: L09501.

The *Washington Post* article on 22 October 2007 is 'At the poles, melting occurring at an alarming rate' by Doug Struck. Regular updates and announcements from the NSIDC are available at nsidc.org/arcticseaice/news. Data on ice thickness was presented by Wieslaw Maslowski to an American Meteorological Society Environmental Science Series Seminar as 'Causes of changes in arctic sea ice' on 3 May 2006.

Walt Meier and Doug Serreze's comments on tipping points were reported in *The Independent* on 15 August 2007 by S. Connor and on 22 September 2007 by M. McCarthy. Connor also reported on Ron Lindsay's hypothesis on 29 December 2006. Ted Scambos' remarks were made in a personal communication on 21 September 2007, and Struck's *Washington Post* article of 22 October 2007. Tim Flannery's comments appeared in *The Age* on 28 October 2006.

Changes in Arctic conditions: W. Maslowski, J. Clement et al. (2006) 'On oceanic forcing of Arctic climate change', *Geophysical Research Abstracts* 8: 05892; D. K. Perovich, J. A. Richter-Menge, et al. (2007) 'Arctic sea ice melt in summer 2007: Surface and bottom ice ablation', Eos Trans. AGU 88(52) Fall Meet. Suppl., Abstract C21C-07; '"Warm wind" hits Arctic climate', BBC News, 18 October 2007.

Sea-ice loss predictions: Louis Fortier – M. White, '"Frightening" projection for Arctic melt', *Ottawa Citizen*, 16 November 2007; Wieslaw Malowski – J. Amos, 'Arctic summers ice-free "by 2013"', BBC News, 12 December 2007; Jay Zwally – A. Beck. 'Arctic's record melt', *Sydney Morning Herald*, 14 December 2007; Josefino Comiso – M. Inman, 'Global warming "tipping points" reached, scientist says', *National Geographic News*, 14 December 2007.

Only 13 per cent of first-year ice survived melt season: 'Arctic sea ice news & analysis', nsidc.org/arcticseaicenews, accessed 17 April 2008. More open water in 2008: Prof. Olav Orheim, Executive Secretary for the International Polar Year Secretariat, personal communication, 10 March 2008.

Albedo flip: J. Hansen (2007) 'Scientific reticence and sea level rise', *Environmental Research Letters* 2: 024002; C. Parkinson, D. Rind et al. (2001) 'The impact of sea ice concentration accuracies on climate model simulations with the GISS GCM', *Journal of Climate* 14: 2606–23. A. Doyle, 'Arctic thaw may be at "tipping point"', Reuters, 28 September 2007. Warmer surface temperatures: S. Borenstein, 'Arctic sea ice gone in summer within five years?', Associated Press, 12 December 2007.

Richard Spinrad's comment was reported by Associated Press, 'Climate change reshaping Arctic', 18 October 2007.

## Chapter 2: Greenland's fate

IPCC Greenland predictions: *Climate Change 2007: synthesis report* (IPCC: Geneva, 2007)

Increased rate of ice mass loss: E. Rignot and P. Kanagaratnam

(2006) 'Changes in the velocity structure of the Greenland ice sheet', *Science* 311: 5763; J. L. Chen, C. R. Wilson et al. (2006) 'Satellite gravity measurements confirm accelerated melting of Greenland ice', *Science* 313: 1958–60; K. Young, 'Greenland ice cap may be melting at triple speed', *New Scientist*, 10 August 2006.

Increased rate of melting: M. Tedesco (2007) 'Snowmelt detection over the Greenland ice sheet from SSM/I brightness temperature daily variations', *Geophysical Research Letters* 34: L02504; T. L. Mote (2007) 'Greenland surface melt trends 1973–2007: Evidence of a large increase in 2007', *Geophysical Research Letters* 34: L22507; E. Saupe, 'Snowmelt on the rise in Greenland', *GeoTimes*, 7 June 2007; D. Shukman, 'Greenland ice-melt "speeding up"', BBC News, 28 July 2007.

Increased air temperature: K. Steffen, R. Huff et al. (2007) 'Arctic warming, Greenland melt and moulins', *Eos Trans. AGU* 88(52) Fall Meet. Suppl., Abstract C21C-07 88(52). Robert Corell's comment was reported by E. Hilton, 'Greenland is now a country fit for broccoli growers', *The Guardian*, 14 September 2007.

Tremors: P. Brown, 'Scientists fear ice caps melting faster than predicted', *The Guardian*, 7 September 2007. Rising landmass: C. Brahic, 'Shrinking ice means Greenland is rising fast', *New Scientist*, 2 November 2007.

Critical melt threshold: J. M. Gregory, P. Huybrechts et al. (2004) 'Climatology: Threatened loss of the Greenland ice-sheet', *Nature* 426: 616; P. Chylek and U. Lohmann (2005) 'Ratio of the Greenland to global temperature change: Comparison of observations and climate modeling results', *Geophysical Research Letters* 32: L14705.

The remarks by Tim Lenton and Lenny Smith were reported by

F. Pearce, 'Climate tipping points loom', *New Scientist*, 16 August 2007. At the threshold: 'Greenland's water loss has doubled in a decade', *New Scientist*, 25 February 2006.

James Hansen's comments: 'Scientific reticence and sea level rise', *Environmental Research Letters* 2: 024002; 'The threat to the planet: Actions required to avert dangerous climate change', presentation at SOLAR 2006 Conference on renewable energy, Denver, 10 July 2006; 'The threat to the planet: How can we avoid dangerous human-made climate change?', remarks on acceptance of WWF Duke of Edinburgh Conservation Medal at St James Palace, 21 November 2006. Hansen's presentations are available at www.columbia.edu/~jeh1/.

Broadening inputs to IPCC reports: M. Oppenheimer, B.C. O'Neill, et al. (2007) 'The limits of consensus', *Science* 317: 1505–06.

Could the ice-sheet survive: J. Hansen, M. Sato, et al. (2007) 'Dangerous human-made interference with climate: a GISS modelE study', *Atmospheric Chemistry and Physics* 7: 2287–312.

Paleoclimate record and sea levels: J. Oerlemans, D. Dahl-Jensen et al., (2006) 'Ice sheets and sea levels', *Science* 313: 1043–45; P. Spotts, 'Little Time to avoid big thaw, scientists warn', *Christian Science Monitor*, 24 March 2006.

The AGU meeting was reported by S. Borenstein, 'Arctic sea ice gone in summer within five years?', Associated Press, 12 December 2007.

## Chapter 3: Trouble in the Antarctic

Sensitivity to warming: S. Schmitt, 'Refrigeration system for the Earth's oceans threatens to break down', *Spiegel online*, 2 March 2007, www.spiegel.de/international/0,1518,469495,00.html

6 degrees since 1950: D. Struck, 'At the poles, melting occurring at an alarming rate', *Washington Post*, 22 October 2007.

Collapse of Larsen B: F. Pearce, *With Speed and Violence: why scientists fear tipping points in climate change* (Beacon Press: Boston, 2007); N. F. Glasser and T. A. Scambos (2008) 'A structural glaciological analysis of the 2002 Larsen B ice shelf collapse', *Journal of Glaciology* 54: 3–16; 'New Research on the 2002 Collapse of the Larsen B Ice Shelf', nsidc.org/news/press/20080207_Scambos.html, accessed 27 February 2008.

The John Mercer story and Pine Island Bay: J. Mercer (1978) 'West Antarctic ice sheet and CO2 greenhouse effect: threat of disaster', *Nature* 271: 321–25; J. Hansen, 'Huge sea level rises are coming—unless we act now', *New Scientist*, 28 July 2007; F. Pearce (as above).

The gravest threat: J. Hansen, M. Sato et al. (2007) 'Climate change and trace gases', *Philosophical Transactions Royal Society* 365: 1925–54

East Antarctica: E. Rignot, J. L. Bamber et al. (2008) 'Recent Antarctic ice mass loss from radar interferometry and regional climate modelling', *Nature Geoscience* 1: 106–110; F. Pearce (as above); M. Tedesco, W. Abdalati et al. (2007), 'Persistent surface snowmelt over Antarctica (1987–2006) from 19.35 GHz brightness temperatures', *Geophysical Research Letters*, 34: L18504; J. E. Francis and Robert S. Hill (1996) 'Fossil plants from the Pliocene Sirius group, Transantarctic Mountains: Evidence for climate from growth rings and fossil leaves', *Palaios*, 11: 389–96.

## Chapter 4: A Rising Tide

The IPCC's sea-level rise predictions in February 2007 are contained in 'Contribution of Working Group I to the Fourth Assessment Report of the Intergovernmental Panel on Climate

Change: Summary for Policymakers'. The final report for 2007 is *Climate Change 2007: synthesis report* (IPCC: Geneva).

Water run-off impounded on land: B. F. Chao, Y. H. Wu et al. (2008) 'Impact of artificial reservoir water impoundment on global sea level', *Science* 320: 212–14.

Under-estimation of rises: S. Rahmstorf, J. Cazenave et al. (2007) 'Recent climate observations compared to projections', *Science* 316: 709; Robert Corell's comment is reported by F. Pearce, 'But here's what they didn't tell us', *New Scientist*, 10 February 2007; M. Oppenheimer, B. C. O'Neill et al. (2007) 'The Limits of Consensus', *Science* 317: 1505–06; P. Brown, 'Scientists fear ice caps melting faster than predicted', *The Guardian*, 7 September 2007.

James Hansen and his colleagues' writings on sea-level rises include 'Scientific reticence and sea level rise', *Environmental Research Letters* 2: 024002 and 'Dangerous human-made interference with climate: a GISS modelE study', Atmospheric Chemistry and Physics 7: 2287–312. The log extract is from 'Huge sea level rises are coming—unless we act now', *New Scientist*, 28 July 2007. Hansen's 1988 remarks are reported 'The public and climate change', www.aip.org/history/climate/public2.htm, accessed 2 March 2008.

The sea-level rise chart is based on D. Archer, *Global Warming: understanding the forecast* (Blackwell Publishers: Oxford, 2006). Hansen's comments at the AGU were reported by A. Beck, 'Carbon cuts a must to halt warming: US scientists', Reuters, 13 December 2007.

Stern's review is *The Economics of Climate Change: the Stern review* (Cambridge: Cambridge, 2006).

Underground water: F. Pearce, 'Cities may be abandoned as salt water invades', *New Scientist*, 16 April 2006

## Chapter 5: The Quickening Pace

Climate sensitivity: N. Andronova and M. E. Schlesinger (2001) 'Objective estimation of the probability distribution for climate sensitivity', *Journal of Geophysical Research* 106: 22605; J. D. Annan and J. C. Hargreaves (2006) 'Using multiple observationally-based constraints to estimate climate sensitivity', *Geophysical Research Letters* 33: L06704; M. Hopkin, 'Climate sensitivity "inherently unpredictable"', *Nature News*, 25 October 2007; B. Pittock (2006) 'Are scientists underestimating climate change?', *Ecos* 87: 34.

'Long run' sensitivity: J. Hansen, 'Can we defuse the global warming time bomb?', *Natural Science*, 1 August 2003; G. H. Roe and M. B. Baker (2007) 'Why is climate sensitivity so unpredictable?', *Science* 318: 629–32; Hansen and M. Sato, 'Global warming: East-West connections' (draft September 2007), www.columbia.edu/~jeh1/East-West_070925.pdf; J. Hansen, M. Sato et al., 'Target atmosphere CO2: Where should humanity aim', submitted to *Science* 7 April 2008, arxiv.org/abs/0804.1126.

Paleoclimate data: M. Scheffer, V. Brovkin et al. (2006) 'Positive feedback between global warming and atmospheric CO2 concentration inferred from past climate change', *Geophysical Research Letters* 33: L10702; A. Sluijs, S. Schouten et al. (2006) 'Subtropical Arctic Ocean temperatures during the Palaeocene/Eocene thermal maximum', *Nature* 441: 61-613

As high as 10 degrees: M. O. Andreae, C. D. Jones et al. (2005) 'Strong present-day aerosol cooling implies a hot future', *Nature* 435: 1187–90

Carbon sinks less effective: C. D. Jones, P. M. Cox et al. (2003) 'Strong carbon cycle feedbacks in a climate model with interactive CO2 and sulphate aerosols', *Geophysical Research*

*Letters* 30: 1479; M. R. Raupach, G. Marland, et al. (2007) 'Global and regional drivers of accelerating CO2 emissions', *Proceedings National Academy Sciences* 104: 10288–93; S. E. Schwartz, R. J. Charlson et al., 'Quantifying climate change — too rosy a picture?', *Nature Reports Climate Change,* 27 June 2007; P. M. Cox, R. A. Betts et al. (2000) 'Acceleration of global warming due to carbon-cycle feedbacks in a coupled climate model', *Nature* 408: 184–87; J. G. Canadell, C. LeQuere et al. (2007) 'Contributions to accelerating atmospheric CO2 growth from economic activity, carbon intensity, and efficiency of natural sinks', *Proceedings National Academy Sciences* 104: 18866–70; F. Pearce, *With Speed and Violence: why scientists fear tipping points in climate change* (Beacon Press: Boston, 2007); P. M. Cox, C. Huntingford et al., 'Conditions for sink-to-source transitions and runaway feedbacks from the land carbon cycle' in H. J. Schellnhuber, W. Cramer at al. (eds) *Avoiding Dangerous Climate Change* (Cambridge: Cambridge, 2006).

Hadley Centre modeling: G. Jenkins, R. Betts, et al., *Stabilising Climate to Avoid Dangerous Climate Change: a summary of relevant research at the Hadley Centre* (Met Office: Exeter, 2005).

Carbon lost by soil and plants: J. Pickrell, 'Soil may spoil UK's climate efforts', *New Scientist,* 7 September 2005; S. Connor and M. McCarthy, 'Our worst fears are exceeded by reality', *The Independent,* 29 December 2006; J. L. Sarmiento and N. Gruber (2003) 'Sinks for anthropogenic carbon', *Physics Today,* August 2002; W. Knorr, W. Gobron et al. (2007) 'Impact of terrestrial biosphere carbon exchanges on the anomalous CO2 increase in 2002–2003', *Geophysical Research Letters* 34: L09703; K. M. Carney, B. A. Hungate et al. (2007) 'Altered soil microbial community at elevated CO2 leads to loss of soil carbon', *Proceedings National Academy Sciences* 104: 4990–95.

Amazon: Y. Malhi, J. T. Roberts et al. (2008) 'Climate change, deafforestation, and fate of the Amazon', *Science* 319: 169–71; R. Butler, 'Amazon rainforest fires 'worst' in memory', Mongabay. com, 16 October 2007, new.mongabay.com/2007/1016-amazon. html, accessed 28 October 2007; D. Howden and J. Steven (2007) 'South America chokes as Amazon burns', *The Independent*, 5 October 2007; T. H. Lenton, H. Held et al. (2008) 'Tipping elements in the Earth's climate system' *Proceedings National Academy Sciences* 105: 1786–93.

Indonesia: S. Page, F. Siegert et al. (2002) 'The amount of carbon released from peat and forest fires in Indonesia during 1997', *Nature* 420: 61–65.

Permafrost: M. T. Jorgenson, Y. L. Shur et al. (2006) 'Abrupt increase in permafrost degradation in Arctic Alaska', *Geophysical Research Letters* 33: L02503; K. M. Walter, S. A. Zimov et al. (2006) 'Melting lakes in Siberia emit greenhouse gas", *Nature* 443: 71–75; D. Struck, 'At the poles, melting occurring at an alarming rate', *Washington Post*, 22 October 2007; D. Solovyov and A. Doyle, 'Siberian thaw could speed up global warming', *Sydney Morning Herald*, 26 September 2007; J. Randerson, 'Forests battle to soak up carbon', *The Age*, 4 January 2008; J. B. Miller (2008) 'Carbon cycle: Sources, sinks and seasons', *Nature* 451: 26–27.

Ocean sinks: C. LeQuere, C. Rodenbeck, et al. (2007) 'Saturation of the southern Ocean CO2 sink due to recent climate change', *Science* 316: 1735–38; A. Woodcock, 'Scientists fear climate change speed-up as oceans fail to hold greenhouse gases', *The Scotsman*, 21 October 2007; M. J. Behrenfeld, K. Worthington et al. (2007) 'Controls on tropical Pacific Ocean productivity revealed through nutrient stress diagnostics', *Nature* 442: 1025–28; J. J. Polovina, E. A. Howell et al. (2008) 'Ocean's least productive waters are expanding', *Geophysical Research Letters* 35: L03618; K. Caldeira, D. Archer et al. (2007) 'Comment on "Modern-age buildup of CO2 and its effects on seawater acidity and salinity" by Hugo A. Loáiciga', *Geophysical Research Letters* 38: L18608; R. Kleinman, 'Warming turns Barrier Reef acidic', *The Age*, 18 October 2007

## Chapter 6: Most Species, Most Ecosystems

Martin Parry's statement was reported by D. Adam, 'How climate change will affect the world, *The Guardian*, 19 September 2007.

Species extinction as temperatures rose 5 degrees: S. T. Jackson and C. Y. Weng (1999) 'Late quaternary extinction of a tree species in eastern North America', *Proceedings National Academy Sciences* 96: 13847–52; S. Rahmstorf, 'Climate change fact sheet', Potsdam Institute for Climate Impact Research, www.pik-potsdam.de/~stefan/warmingfacts.pdf.

IPCC worst-case scenario: C. H. Sekercioglu, S. H. Schneider et al. (2008) 'Climate change, elevational range shifts, and bird extinctions', *Conservation Biology* 22: 140–50; E. Marris, 'The escalator effect', *Nature reports climate change*, 23 November 2007.

Rate of change is analysed by R. Leemans and B. Eickhout (2004) 'Another reason for concern: regional and global impacts on ecosystems for different levels of climate change', *Global Environmental Change* 14: 219–28.

A 2007 study is J. W. Williams, S. T. Jackson et al. (2007) 'Projected distributions of novel and disappearing climates by 2100 AD', *Proceedings National Academy Sciences* 104: 5738–42.

1350 European plant species: W. Thuiller, S. Lavorel et al. (2005) 'Climate change threats to plant diversity in Europe', *Proceedings National Academy Sciences* 102: 8245–50

Over the last 25 years: D. J. Seidel Q. Fu et al. (2008) 'Widening
of the tropical belt in a changing climate', *Nature Geoscience* 1:
21–24; D. J. Seidel and W. J. Randel (2007) 'Recent widening
of the tropical belt: Evidence from tropopause observations',
*Journal Geophysical Research* 112: D20113; S. Connor, 'Expanding
tropics "a threat to millions"', *The Independent*, 3 December
2007.

If the rate should exceed 4 degrees: S. Kallbekken and J. S.
Fuglestvedt, 'Faster change means bigger problems', Centre
for International Climate and Environmental Research – Oslo,
www.cicero.uio.no/fulltext/index_e.aspx?id=5690, accessed 24
December 2007.

## Chapter 7: The Price of Reticence

Roger Jones' *Herald Sun* article on 10 December 2007 is 'Keep
cool is the message on climate'.

Serious disagreement: R. McKie, 'Experts split over climate
danger to Antarctica', *The Observer*, 28 January 2007; F. Pearce,
'But here's what they didn't tell us', *New Scientist*, 10 February
2007.

Uncertainties in climate science: B. Pittock (2006) 'Are scientists
underestimating climate change?', *Ecos* 87: 34; M. Oppenheimer,
B. C. O'Neill et al. (2007) 'The Limits of Consensus', *Science* 317:
1505–06; J. Hansen and M. Sato (2007) 'Global warming: East-
West connections', draft of September 2007, www.columbia.
edu/~jeh1/2007/EastWest_20070925.pdf.

Limitations of IPCC process: B. Pittock, 'Ten reasons why
climate change may be more severe than projected' in M. C.
MacCracken, F. Moore et al. (eds), *Sudden and Disruptive Climate
Change* (Earthscan: London, 2008).

Inez Fung: C. Barras, 'Rocketing CO2 prompts criticisms of

IPCC', *New Scientist*, 24 October 2007.

Will Steffen's comments are reported by F. Pearce in *With Speed and Violence: why scientists fear tipping points in climate change* (Beacon Press: Boston, 2007).

## Chapter 8: What We Are Doing

Greenhouse gases: *Climate Change 2007: synthesis report. A report of the Intergovernmental Panel on Climate Change* (WMO/ UNEP, 2007); J. Houghton, *Global Warming: the complete briefing* (Cambridge: Cambridge, 2004); J. Hansen, M. Sato et al. (2007) 'Climate change and trace gases', *Philosophical Transactions Royal Society* 365: 1925–54

Carbon dioxide: 'The global carbon cycle', UNESCO/Scope Policy Brief No. 2, October 2006; H. D. Matthews and K. Caldeira (2008) 'Stabilizing climate requires near-zero emissions', Geophysical Research Letters 35: L04705.

Methane: D. T. Shindell, G. Faluvegi et al. (2005) 'An emissions-based view of climate forcing by methane and tropospheric ozone', *Geophysical Research Letters* 32: L04803.

Aerosols: N. Bellouin, O. Boucher et al. (2005) 'Global estimate of aerosol direct radiative forcing from satellite measurements', *Nature* 438: 1138–41; M. O. Andreae, C. D. Jones et al. (2005) 'Strong present-day aerosol cooling implies a hot future', *Nature* 435: 1187–90; S. E. Schwartz, R. J. Charlson et al., 'Quantifying climate change — too rosy a picture?', *Nature Reports: Climate Change*, 27 June 2007; 'Faster climate change predicted as air quality improves', Max Plack Institute for Chemistry press release, 30 June 2005, www.mpch-mainz.mpg.de/mpg/english/ pri0805.htm; V. Ramanathan and G. Carmichael (2008) 'Global and regional climate changes due to black carbon', *Nature GeoScience* 1: 221–27.

## Chapter 9: Where We Are Headed

IPCC scenarios are available at www.ipcc.ch/ipccreports/sres/ emission. Roger Jones's comment is 'Keep cool is the message on climate', *Herald Sun*, 10 December 2007.

Carbon dioxide emissions: M. R. Raupach, G. Marland et al. (2007) 'Global and regional drivers of accelerating $CO_2$ emissions', *Proceedings National Academy Sciences* 104: 10288–93; '$CO_2$ emissions increasing faster than expected', CSIRO media release 07/89, 22 May 2007, www.csiro.au/news/GlobalCarbonProject–PNAS.html; R. W. Bacon and S. Bhattacharya, *Growth and CO2 Emissions: how do different countries fare* (The World Bank Environment Department: Washington, 2007).

Carbon dioxide concentrations: A. Arguez, A. M. Waple et al. (2007) 'State of the climate in 2006', *Bulletin of the American Meteorological Society* 88: 929–32; 'Carbon dioxide, methane rise sharply in 2007', National Oceanic and Atmospheric Administration, 23 April 2008, www.noaanews.noaa.gov/stories2008/20080423_methane.html; S. Connor, 'If we fail to act, we will end up with a different planet', *The Independent*, 1 January 2007; J. Amos, 'Deep ice tells long climate story', BBC News, 4 September 2006.

Energy use: The International Energy Agency's annual *World Energy Outlook* reports are available at worldenergyoutlook.org; *World Energy Technology Outlook – 2007* (European Commission: Brussels, 2007); D. Parsely, 'Climate change "is accelerating"', *The Observer*, 23 March 2008.

Rising temperatures: S. Rahmstorf, J. Cazenave et al. (2007) 'Recent climate observations compared to projections', *Science* 316: 709; D. M. Smith, S. Cusack et al. (2007) 'Improved surface temperature prediction for the coming decade from a global

climate model', *Science* 317: 796–99; J. Hansen, M. Sato et
al. (2006) 'Global temperature change', *Proceedings National
Academy Sciences* 103: 14288–93.

Tipping points: J. von Radowitz, 'Ice cores hold threat of
climate timebomb', *The Age*, 6 September 2006; J. Hansen,
'The threat to the planet: how can we avoid dangerous human-
made climate change?', remarks on acceptance of WWF Duke
of Edinburgh Conservation Medal at St. James Palace, 21
November 2006; W. Steffen, 'Climate Change: science, impacts
and policy challenges', *Policy Briefs 5* (Crawford School of
Economics and Government, ANU: Canberra, 2007); J. Hansen,
private communication, 29 March 2007; T. Colebatch, 'The
European solution', *The Age*, 24 October 2006.

## Chapter 10: Target 2 Degrees

Sir John Holmes: J. Borger, 'Climate change disaster is upon us,
warns UN', *The Guardian*, 5 October 2007

1-degree and 2-degree impacts are surveyed by M. Lynas, *Six
Degrees: our future on a hotter planet* (Fourth Estate: London,
2007). North Queensland: J. W. Williams, S. T. Jackson et
al. (2007) 'Projected distributions of novel and disappearing
climates by 2100 AD', *Proceedings National Academy Sciences* 104:
5738–42; L. Minchin, 'Reef "facing extinction"', *The Age*, 30
January 2007.

Setting goals: 'The Toronto and Ottawa conferences and the
"Law of the atmosphere"', geography.otago.ac.nz/Mirrors/
Climatechange-Factsheets_Mirror/fs215.html; I. Enting, T.
Wigley et al., *Future Emissions and Concentrations of Carbon
Dioxide: key ocean/atmosphere/land analyses*, Technical Paper No.
31 (CSIRO Division of Atmospheric Research: Melbourne, 1994);
J. Leggett, *The Carbon War: global warming and the end of the oil era*

(Routledge: New York, 2001); N. Stern, *The Economics of Climate Change: the Stern review* (Cambridge: Cambridge, 2006).

2-degree scenarios: M. Meinshausen, 'What does a 2°C target mean for greenhouse gas concentrations? A brief analysis based on multi-gas emission pathways and several climate sensitivity uncertainty estimates' in H. J. Schellnhuber, W. Cramer at al. (eds) *Avoiding Dangerous Climate Change* (Cambridge: Cambridge, 2006); M. Meinshausen, 'Less than 2°C trajectories: a brief background note', KyotoPlus conference papers, 28–29 September 2006, Berlin; S. Rettalack, *Setting a Long-term Climate Objective: a paper for the International Climate Change Taskforce* (Institute for Public Policy Research: London, 2005); P. Baer and M. Mastrandrea, *High Stakes: designing emissions pathways to reduce the risk of dangerous climate change* (Institute for Public Policy Research: London, 2006); N. Rive, A. Torvanger et al. (2007) 'To what extent can a long-term temperature target guide near-term climate change commitments', *Climatic Change* 82: 373–91.

After a careful reassessment: J. Hansen, M. Sato et al., 'Target atmosphere CO2: Where should humanity aim', submitted to *Science* 7 April 2008, arxiv.org/abs/0804.1126.

## Chapter 11: Getting The Third Degree

Graeme Pearman's commentary is 'Bali—high urgency', ABC online opinion, 3 December 2007, www.abc.net.au/unleashed/stories/s2108079.htm

3-degree impacts are surveyed by M. Lynas, *Six Degrees: our future on a hotter planet* (Fourth Estate: London, 2007).

Labor policy: P. Garrett, 'Labor's greenhouse reduction target: 60% by 2050 backed by the science', Media statement, 2 May 2007; Penny Wong: 'Aust "most vulnerable" to climate change:

Garnaut', ABC News, 21 February 2008.

The 60/2050 goal: The UK Royal Commission on Environmental Pollution's 2000 report *Energy: the changing climate* is available at www.rcep.org.uk/energy.htm; J. Leggett, *The Carbon War: global warming and the end of the oil era* (Routledge: New York, 2001).

Stern's target: N. Stern, 'Launch presentation', 30 October 2006 and 'Executive summary: Stern Review on the Economics of Climate Change', October 2006, www.hm-treasury.gov. uk/independent_reviews/stern_review_economics_climate_change; P. Hannam, 'New Stern climate warning', *The Age*, 28 March 2007.

An ice-free planet: J. Hansen, M. Sato et al., 'Target atmosphere CO2: Where should humanity aim', submitted to *Science* 7 April 2008, arxiv.org/abs/0804.1126.

The British government: G. Monbiot, 'Giving up on two degrees', *The Guardian*, 1 May 2007.

The CSIRO report is R. N. Jones and B. L. Preston, *Climate Change Impacts, Risk and the Benefits of Mitigation: a report for the Energy Futures Forum* (CSIRO Marine and Atmospheric Research: Melbourne, 2006).

James Lovelock's analysis is described in *The Revenge of Gaia* (Allen Lane: London, 2006); J. Lovelock and L. R. Kump (1994) 'Failure of climate regulation in a geophysiological model', *Nature* 369: 732–34.

Stern's 2008 comments: 'Climate expert Stern "underestimated problem"', *The Age*, 17 April 2008.

We'd all vote to stop climate change: P. Baer and T. Athanasiou, 'Honesty about dangerous climate change', www.ecoequity. org/ceo/ceo_8_2.htm, accessed 3 January 2007.

## Chapter 12: Planning The Alternative

Nicolas Stern's 4 January 2008 presentation is 'The economics of climate change', Richard T. Ely Lecture, AEA Meeting, New Orleans, www.occ.gov.uk/activities/stern_papers/Ely lecture 20.12.2007 no notes.pdf.

Scale similar to wars and Depression: N. Stern, 'Executive summary: Stern Review on the Economics of Climate Change', October 2006.

Tony Coleman's comments were reported by K. Davidson, 'Being the "lucky country" will not save us from climate change', *The Age*, 20 March 2008.

UNFCCC objective: unfccc.int/essential_background/ convention/background/items/1353.php

## Chapter 13: The Safe-Climate Zone

Overviews of paleoclimatology, the study of the Earth's climate history, are available at a number of web portals, including ncdc. noaa.gov/paleo/paleo.html and earthobservatory.nasa.gov/ Study/Paleoclimatology.

Sea levels and the Holocene: J. Hansen (2005) 'A slippery slope: how much global warming constitutes "dangerous anthropogenic interference"? An editorial essay', *Climate Change* 68: 269–79; J. Hansen (2007) 'Scientific reticence and sea level rise', *Environmental Research Letters* 2: 024002

Hansen threshold: J. Hansen, M. Sato et al., 'Target atmosphere $CO_2$: Where should humanity aim', submitted to *Science* 7 April 2008, arxiv.org/abs/0804.1126.

Sea-ice flushing: M. Serreze, M. Holland et al. (2007) 'Perspectives on the Arctic's shrinking sea ice cover', *Science* 315: 1533–36; 'Arctic sea ice decline may trigger climate change cascade', University of Colorado media release, March 15, 2007, www.colorado.edu/news/releases/2007/109.html; Y. Yu, G. A. Maykut et al. (2004) 'Changes in the thickness distribution of Arctic sea ice between 1958–1970 and 1993–1997', *Journal Geophysical Research* 109: C08004.

Insight from early Holocene: S. Funder and K. H. Kjaer (2007) 'Ice free Arctic Ocean, an Early Holocene analogue', Eos Trans. AGU, 88(52), Fall Meet. Suppl., Abstract PP11A-0203; F. Wagner, B. Aaby et al. (2002) 'Rapid atmospheric CO2 changes associated with the 8,200-years-B.P. cooling event', *Proceedings National Academy Sciences* 99: 12011–14.

Reducing atmospheric carbon: H. D. Matthews and K. Caldeira (2008) 'Stabilizing climate requires near-zero emissions', *Geophysical Research Letters* 35: L04705.

Cutting emissions to zero: T. Helweg-Larsen and J. Bull, *Zero Carbon Britain: an alternative energy strategy* (Centre for Alternative Technology, 2007);

Aerosols dilemma and direct cooling: Committee on Science, Engineering, and Public Policy (National Academy of Sciences, National Academy of Engineering, Institute of Medicine), *Policy Implications of Greenhouse Warming: mitigation, adaptation, and the science base* (National Academy Press: Washington, DC, 1992); P. J. Crutzen (2006) 'Albedo enhancement by stratospheric sulfur injections: a contribution to resolve a policy dilemma? Editorial Essay', *Climatic Change* 77: 211–19; H. D. Matthews and K. Caldeira (2007) 'Transient climate-carbon simulations of planetary geoengineering', *Proceedings National Academy Sciences* 104: 9949–54; M. O. Andreae, C. D. Jones et al. (2005) 'Strong present-day aerosol cooling implies a hot future', *Nature* 435: 1187–90; Q. Schiermeier (2007) 'Climate change 2007: what we don't know about climate change', *Nature* 445: 580–81.

## Chapter 14: Putting The Plan Together

Momentous political tipping point: M. Inman, 'Global warming "tipping points" reached, scientist says', National Geographic News, 14 December 2007; A. Beck, 'Carbon cuts a must to halt warming—US scientists', Reuters, 13 December 2007; S. Borenstein, 'Arctic sea ice gone in summer within five years?', Associated Press, 12 December 2007.

Restoration of Arctic sea-ice: J. Hansen, M. Sato et al., 'Target atmosphere CO2: where should humanity aim', submitted to *Science* 7 April 2008, arxiv.org/abs/0804.1126.

Less than 1.7 degrees: J. Hansen (2005) 'A slippery slope: how much global warming constitutes "dangerous anthropogenic interference"? An editorial essay', *Climate Change* 68: 269–79; J. Hansen, 'Tipping point: perspective of a climatologist' in E. Fearn and K. H. Redford (eds) *The State of the Wild 2008: a global portrait of wildlife, wildlands and oceans* (Wildlife Conservation Society/Island Press, 2008). Court testimony: J. Hansen (2007) 'Direct testimony of James E. Hansen', State of Iowa before the Iowa Utilities Board, docket no. GCU-07-1, 22 October 2007, www.columbia.edu/~jeh1/IowaCoal_071105.pdf

Lehman Brothers report: J. Llewellyn, J. and C. Chaix 'The business of climate change II: Policy is accelerating, with major implications for companies and investors", Lehman Brothers, 20 September 2007, www.lehman.com/who/intcapital/#

Greenhouse development rights: P. Baer, T. Athanasiou and S. Kartha (2007) 'The Right to Development in a Climate Constrained World: The Greenhouse Development Rights Framework', EcoEquity, www.ecoequity.org/GDRs/.

## Chapter 15: This Is An Emergency

Ban Ki-Moon statement: ABC News, 'UN chief says global

warming is 'an emergency'', 11 November 2007. The Apollo metaphor was suggested by David Wasdell of the Meridian Programme.

The era of catastrophic climate change: James Hansen raised this subject in 2005 in 'A slippery slope: how much global warming constitutes "dangerous anthropogenic interference"? An editorial essay', *Climate Change* 68: 269–79 and in subsequent presentations including 'The threat to the planet: actions required to avert dangerous climate change' at the Solar Conference on Renewable Energy in Denver on 10 July 2006.

## Chapter 16: A Systemic Breakdown

The Queensland Government April 2007 report is *Queensland's Vulnerability to Rising Oil Prices: Taskforce report*. Economic restructuring to solve the peak oil crisis is surveyed in Robert Hirsch, Roger Bezdek, et al., *Peaking of World Oil Production: impacts, mitigation, and risk management* (Novinka Books: New York, 2005). Oil demand growth: J. Mouawad, 'Cuts urged in China's and India's energy growth', *New York Times*, 7 November 2007; M. T. Klare, 'The bad news at the pump: the $100-plus barrel of oil and what it means', TomDispatch.com, 11 March 2008, www.tomdispatch.com/post/174904.

The contamination of groundwater is reported by Fred Pearce, *When the Rivers Run Dry: water—the defining crisis of the twenty-first century* (Beacon Press: Boston, 2007). Land use: United Nations Environment Program, *Global Environment Outlook: Environment for development* (GEO-4), (Progress Press: Malta, 2007).

Food prices and biofuels: Peter Weekes, 'Storm clouds on the horizon', *The Age*, 22 December 2007; Javier Blas, Chris Giles and Hal Weitzman, 'World food price rises to hit

consumers', *Financial Times*, 16 December 2007. Brittany
Sauser, 'Ethanol demand threatens food prices, *MIT Technology
Review*, 13 February 2007; Geoffrey Styles, 'The biofuel gap',
energyoutlook.blogspot.com/2008/01/biofuel-gap.html,
accessed 7 January 2008; D. Nason, 'First signs of the coming
famine', *The Weekend Australian*, 26–27 April 2008.

### Chapter 17: When 'Reasonable' Is Not Enough

Extensive ecosystem damage: F. J. Rijsberman and R. J. Swart
(eds.), *Targets and Indicators of Climate Change* (Stockholm
Environment Institute: Stockholm, 1990); W. K. Stevens, 'Earlier
harm seen in global warming', *The New York Times*, 17 October
1990.

Stern's advocacy of 550pmm/3-degree target may be found
in Part 2.6 of his 2006 'Review on the economics of climate
change' for the UK Treasury. An analysis of Labor's 3-degree
advocacy is provided in Chapter 11. The IPCC 'Summary for
Policymakers of the Synthesis Report of the IPCC Fourth
Assessment Report' released in November 2007, in Table SPM.6,
provides no modeling below the target range 2–2.4 degrees.

ACF climate change campaigner Tony Mohr called for
'emissions to 60–90 per cent below 1990 levels by 2050',
according to AAP, 'Taskforce must aim for radical emissions
cuts: ACF', 6 February 2007. The Garnaut Climate Change
Review's *Interim report* was released on 21 February 2008.

Ken Ward's comments: 'The chasm between our agenda and
climate science – The problem statement: It's time to accept dire
climate realities', 18 April 2007, gristmill.grist.org/story/2007/
4/18/111843/339; 'Bright lines: An introduction – A new path
forward for climate change campaigners', Gristmill, 7 February
2007, gristmill.grist.org/story/2007/2/6/171750/4623.

## Chapter 18: The Gap Between Knowing and Not Knowing

Guy Pearce's book is *High and Dry* (Penguin, 2007). George Monbiot's comments appeared in 'What is progress', *The Guardian*, 4 December 2007 and 'The road well travelled', *The Guardian*, 30 October 2007.

Stanley Cohen's 2001 book is *States of Denial: knowing about atrocities and suffering* (Polity: Boston). Max Bazerman's 2006 essay 'Climate change as a predictable surprise' appeared in *Climatic Change* 77: 179–93. George Marshall comments are from 'The psychology of denial: our failure to act against climate change', *The Ecologist online*, 22 September 2001.

Max Planck's statement is from *Scientific Autobiography* (Philosophical Library: New York, 1949).

## Chapter 19: Making Effective Decisions

Churchill's comments are reported in G. Best, *Churchill: a study in greatness* (Penguin: London, 2002). Jim Collins book is *Good to Great: why some companies make the leap and others don't* (Random House: London, 2001).

John Howard's comments were made on ABC TV's *Lateline*, 5 February 2007, www.abc.net.au/lateline/content/2006/s1840963.htm.

## Chapter 20: The 'New Business–as–usual'

Clean coal: Labor's statement is 'Federal Labor's $500 million national clean coal initiative', www.alp.org.au/media/0207/ms250.php. The 2007 MIT report is *The Future of Coal: options for a carbon–constrained world* and the 2005 IPCC special report is *Carbon Dioxide Capture and Storage*. R. Beeby reported on the parliamentary inquiry, 'Carbon capture costs earth: scientists', *The Canberra Times*, 13 September 2006. The report on *Concentrating Solar Thermal for Australia* was produced by the Cooperative Research Centre for Coal in Sustainable Development. For a comparison of future energy generation costs: *Renewable Energy: a contribution to Australia's environmental and economic sustainability, Final Report to Renewable Energy Generators Australia* by McLennan Magasanik Associates.

Biofuels: Reports on biofuels were published in January 2008 by The Royal Society, *Sustainable Biofuels: prospects and challenges*, and the British House of Commons, *Are biofuels Sustainable?* The 2007 UN report is *The Impact of Biofuels on the Right to Food*. The effect of nitrogen release from biofuel cropping on global warming is demonstrated in P. Crutzen, A. Mosier et al. (2007) 'N2O release from agro-biofuel production negates global warming reduction by replacing fossil fuels', *Atmospheric Chemistry and Physics Discus*sion 7: 11191–205. Energy efficiency comparison is discussed by K. Kleiner in 'The backlash against biofuels', *Nature Reports Climate Change*, January 2008. Biofuel impact on food production: J. Vidal, 'The looming food crisis', *The Guardian*, 29 August 2007 and G. Monbiot, 'An agricultural crime against humanity', *The Guardian*, 6 November 2007

Offsets: The study of effect of tree coverage on climate is S. Gibbard, K. Caldeira et al. (2005) 'Climate effects of global land cover change', *Geophysical Research Letters* 32: L23705. The treddle pump story is told by D. Kennedy and A. O'Connor, 'To cancel out the CO2 of a return flight to India, it will take one poor villager three years of pumping water by foot. So is carbon offsetting the best way to ease your conscience?', *The Times*, 28 August 2007. The *Financial Times* investigation 'Industry caught in carbon smokescreen' by F. Harvey and S. Fidler appeared on 25 April 2007.

Carbon trading and CDMs: A comprehensive study is 'Carbon trading: a critical conversation on climate change, privatization and power', *Development Dialogue* No. 48, September 2006 (The Dag Hammarskjold Foundation/Corner House). *The Guardian* investigation by N. Davies was published on 2 June 2007. The *Newsweek* story 'Global warming: no easy fix' by E. F. Vencat appeared on 12 March 2007. G. Lipow discusses market and regulatory approaches at 'Emissions trading: a mixed record, with plenty of failures', Gristmill, 19 February 2007, http://gristmill.grist.org/story/2007/2/18/205116/813.

## Chapter 21: Climate Solutions

Cutting emissions to zero: T. Helweg-Larsen and J. Bull, *Zero Carbon Britain: an alternative energy strategy* (Centre for Alternative Technology, 2007); A. Makhijani, *Carbon-free and Nuclear-free: a roadmap for US energy policy* (IEER Press/RDR Books, 2007).

Using telecommunications networks: K. Mallon, G. Johnston, et al., *Towards a High-bandwidth, Low-carbon Future: telecommunications-based opportunities to reduce greenhouse gas emissions* (Climate Risk: Sydney, 2007).

McKinsey reports: E. Per-Anders, T. Naucler et al., 'A cost curve for greenhouse gas reduction', *McKinsey Quarterly* 2007(1): 35–45; A. Lewis, S. Gorner et al., *An Australian Cost Curve for Greenhouse Gas Reduction* (McKinsey&Company: Sydney, 2008).

Energy efficiency improvement: L. Brown, *Plan B 3.0: mobilizing to save civilization* (W. W. Norton & Company: New York, 2008). Zero emission homes: H. Osborne, 'First zero-carbon home unveiled', *The Guardian*, 11 June 2007; 'Code for sustainable homes', Planning Portal, www.planningportal.gov.uk/england/professionals/en/1115314116927.html.

Baseload renewable energy: M. Diesendorf, *Greenhouse Solutions with Sustainable Energy* (University of New South Wales Press: Sydney, 2007); M. Diesendorf, 'The base load fallacy', 2007, www.sustainabilitycentre.com.au/BaseloadFallacy.pdf

Generating costs: *Renewable Energy: a contribution to Australia's environmental and economic sustainability: final report to Renewable Energy Generators Australia* (McLennan Magasanik Associates, Melbourne, 2006); R. Roy and G. Mawer, *Putting Renewables on Target: a 10% mandatory renewable energy target* (Greenpeace Australia: Sydney, 2002); K. Mallon and J. Reardon *Cost Convergence of Wind Power and Conventional Generation in Australia: a report for the Australian Wind Energy Association* (AusWEA: Melbourne, 2004); T. Bradford, *The Economic Transformation of the Global Energy Industry* (MIT Press: Cambridge MA, 2006).

Solar thermal: L. Wibberley, A. Cottrell et al., *Concentrating Solar Thermal for Australia*, (Cooperative Research Centre for Coal in Sustainable Development: Pullenvale, 2005); 'Clean power from deserts', TREC/Club of Rome, www.desertec.org.

Materials production: P. Weaver, L. Jansen at al., *Sustainable Technology Development* (Greenleaf Publishing: Sheffield, 2000).

Terra preta: E. Marris gives an overview in 'Black is the new green', *Nature* 442: 624–26. Joe Herbertson's comments are reported in K. Wilson, 'Birth of a new wedge', Truthout, 3 May 2007, http://www.truthout.org/docs_2006/050307R.shtml. Glaser's 2007 paper is 'Prehistorically modified soils of central Amazonia: a model for sustainable agriculture in the twenty-first century', *Philosophical Transactions Royal Society* 362: 187–96. Studies on biochar potential include M. Obersteiner, S. Nilsson. et al., *Biomass Energy, Carbon Removal and Permanent Sequestration: a 'real option' for managing climate risk* (International Institute for

Applied Systems Analysis Interim Report IR-02-042, 2002); J. S. Rhodes and D. W. Keith (2005) 'Engineering economic analysis of biomass IGCC with carbon capture and storage', *Biomass and Energy* 29: 440–50; C. Azar, K. Lindgren et al. (2006) 'Carbon capture and storage from fossil fuels and biomass — Costs and potential role in stabilizing the atmosphere', *Climatic Change* 74: 47–79; and J. Lehmann, J. Gaunt et al. (2006) 'Bio-char sequestration in terrestrial ecosystems – a review', *Mitigation and Adaptation Strategies for Global Change* 11: 403–27.

## Chapter 22: Can 'Politics As Usual' Solve The Problem?

Lovelock's comment is from his 2006 book *The Revenge of Gaia* (Allen Lane: London). Fuel prices and trade: T. Homer-Dixon, *The Upside of Down* (Text: Melbourne, 2006).

Robert Reich's essay, 'How capitalism is killing democracy' appeared in *Foreign Policy*, September/October 2007. See also his book *Supercapitalism: the transformation of business, democracy, and everday life* (Scribe: Melbourne, 2008). Ian Dunlop's comments on 'Corporate governance and responsibility' were made at a book launch at University of Technology Sydney on 5 December 2007.

## Chapter 23: What Does An Emergency Look Like?

Lester Brown's book is *Plan B 3.0: mobilizing to save civilization* (W. W. Norton & Company: New York, 2008).

## Chapter 24: The Climate Emergency In Practice

J. P. Lederach's 1997 book is *Building Peace: sustainable reconciliation in divided societies* (United States Institute of Peace Press: Washington).

Climate impacts: M. Lynas, *Six Degrees: our future on a hotter planet* (Fourth Estate: London, 2007); F. Pearce, *With Speed and Violence: why scientists fear tipping points in climate change* (Beacon Press: Boston, 2007). Kim Stanley Robinson's climate change trilogy is published by Bantam Books, New York.

Watermark Australia's report *Our Water Mark: Australians making a difference in water reform* is available at www. watermarkaustralia.org.au.

Netherland's program: P. Weaver, L. Jansen et al., *Sustainable Technology Development* (Greenleaf Publishing: Sheffield, 2000).

### Chapter 25: The Safe-Climate Economy

Effect of aircraft emissions: J. E. Penner, D. H. Lister et al. (eds) *Aviation and the Global Atmosphere* (Cambridge University Press: Cambridge, 1999), especially chapter 6; G. Monbiot, 'Preparing for take-off', *The Guardian*, 19 December 2006.

The experience of war rationing: Mark Roodhouse, 'Rationing returns: a solution to global warming?', *History and Policy Journal*, March 2007.

Carbon rationing and personal carbon allowances: G. Monbiot, *Heat: how to stop the planet burning* (London: Allen Lane, 2006); R. Starkey and K. Anderson, *Domestic Tradable Quotas: a policy instrument for reducing greenhouse gas emissions from energy use* (Tyndall Centre for Climate Change Research, 2005); S. Roberts and J. Thumim, *A Rough Guide to Individual Carbon Trading: the ideas, the issues and the next steps: report to Defra* (Centre for Sustainable Energy/DEFRA, 2006); 'Domestic tradable quotas as an alternative to carbon taxation: a submission by Feasta to the Department of Finance', October 2003, www.feasta.org/ documents/energy/dtqsoct2003.htm; D. Fleming, 'Energy and the common purpose: descending the energy staircase with tradable energy quotas ('TEQs)', www.teqs.net.

D. Miliband's comments: 'The great stink: towards an
environmental contract', Audit Commission annual lecture, 19
July 2006, www.defra.gov.uk/corporate/ministers/speeches/
david-miliband/dm060719.htm.

## Chapter 26: In the End

Yvo de Boer's comments were report by *The Age*, 'PM flies into
Bali row', 11 December 2007.

Military expenditure 1939–44: Mark Harrison, *The Economics
of World War II: six great powers in international comparison*
(Cambridge: Cambridge, 2000).

# Index